D1564859

ANALYTIC GEOMETRY

*the text of this book is printed
on 100% recycled paper*

COLLEGE OUTLINE SERIES

ANALYTIC GEOMETRY

BY

C. O. OAKLEY, Ph.D.

Professor Emeritus, Department of Mathematics
Haverford College

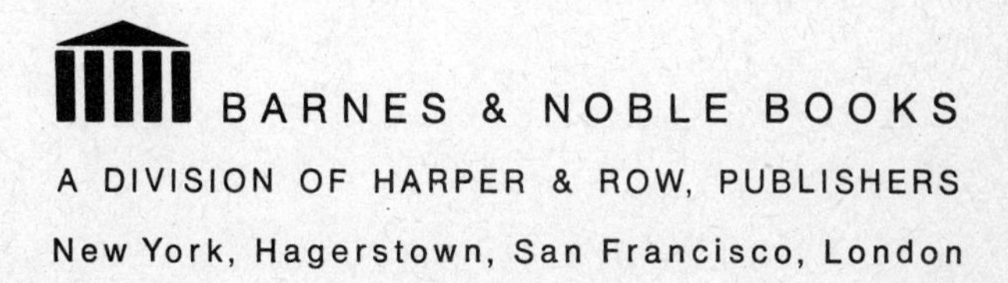

L. C. catalogue card number: 57-12291

ISBN: 0-06-460068-8

79 80 12 11 10 9 8 7

Printed in the United States of America

PREFACE

The principal objective of this Outline is to cover in condensed form suitable for self-instruction and review the subject matter of a first course in Analytic Geometry. To this end we have treated a wide variety of topics in both two and three dimensions; though the list is by no means exhaustive, few courses will include all of this material.

Just as a knowledge of Analytic Geometry is necessary to the study of the Calculus, so certain studies in Algebra, Plane and Solid Geometry, and Trigonometry are essential preliminaries to Analytic Geometry. Chapter I is devoted wholly to basic review and reference formulae in these latter fields.

Many proofs of theorems are included in this Outline in order to satisfy the natural desire of the serious student to know how certain formulae are derived. These derivations, the carefully worked-out Illustrations, and the more than 200 accurately drawn figures should be of material aid to the person who seeks to gain a basic understanding of the processes involved. Typical and standard problems in the form of Exercises, for which answers are supplied, are inserted at the end of each topic. The sample examinations given in Appendix A should help in preparing for quizzes and final tests. Appendix B contains some useful tables.

C. O. O.

Haverford, Pennsylvania

TABLE OF CONTENTS

CHAPTER I

Reference Formulae

PLANE ANALYTIC GEOMETRY

CHAPTER II

Fundamental Concepts

CHAPTER III

Equations and Graphs

CHAPTER IV

The Straight Line

CHAPTER V

The Circle

CHAPTER VI

The Parabola

CHAPTER VII

The Ellipse

CHAPTER VIII

The Hyperbola

CHAPTER IX

Conic Sections

CHAPTER X

Transformations of Coordinates

CHAPTER XI

General Equation of Second Degree

CHAPTER XII

Poles and Polars

CHAPTER XIII

Diameters

CHAPTER XIV

Polar Coordinates

CHAPTER XV

Higher Plane Curves

CHAPTER XVI

Parametric Equations

CHAPTER XVII

Empirical Equations

SOLID ANALYTIC GEOMETRY

CHAPTER XVIII

Fundamental Concepts

CHAPTER XIX

The Plane

CHAPTER XX

The Straight Line

CHAPTER XXI

Space Loci

CHAPTER XXII

The Quadric Surface

APPENDICES

Appendix A

Appendix B

About the Author

C. O. Oakley received the degrees of B.S. in engineering at the University of Texas, M.S. in mathematics at Brown University, and Ph.D. in mathematics at the University of Illinois. He has taught in all these institutions as well as in Bryn Mawr College and the University of Delaware. He is now Professor Emeritus of mathematics at Haverford College. Professor Oakley is a Governor of the Mathematical Association of America, and has served as general mathematics editor of *Colliers Encyclopedia*. His publications include more than twenty mathematical papers in American and foreign learned journals, *Principles of Mathematics* (with C. B. Allendoerfer), and *The Calculus and Analytic Geometry Problems* (both in the College Outline Series).

GREEK ALPHABET

Letters		Names	Letters		Names	Letters		Names
Α	α	Alpha	Ι	ι	Iota	Ρ	ρ	Rho
Β	β	Beta	Κ	κ	Kappa	Σ	σ	Sigma
Γ	γ	Gamma	Λ	λ	Lambda	Τ	τ	Tau
Δ	δ	Delta	Μ	μ	Mu	Υ	υ	Upsilon
Ε	ϵ	Epsilon	Ν	ν	Nu	Φ	ϕ	Phi
Ζ	ζ	Zeta	Ξ	ξ	Xi	Χ	χ	Chi
Η	η	Eta	Ο	ο	Omicron	Ψ	ψ	Psi
Θ	θ	Theta	Π	π	Pi	Ω	ω	Omega

ANALYTIC GEOMETRY

CHAPTER I

REFERENCE FORMULAE

1. Basic Formulae. The student will find it desirable to have before him certain reference formulae which he may consult from time to time. We therefore begin this Outline with a selection of the more important formulae taken from the fields of algebra, geometry, and trigonometry. A thorough review of these before going on to the study of analytic geometry will be of great aid to the student.

2. Algebra.

(1) *Quadratic equation.* The roots (solutions) of the quadratic equation $ax^2 + bx + c = 0$ are

$$x = \frac{-b \pm \sqrt{b^2 - 4\,ac}}{2\,a}.$$

The expression $\Delta = b^2 - 4\,ac$ is called the discriminant.

(a) If $\Delta > 0$, the roots are real and distinct;
(b) If $\Delta = 0$, the roots are real and equal;
(c) If $\Delta < 0$, the roots are complex.

(2) *Factorial notation.* The symbol $n!$, called "n factorial," stands for the product of the first n (positive) integers.

(a) $n! = 1 \cdot 2 \cdot 3 \cdots n$; (b) $0! = 1$, by definition.

(3) *Binomial theorem.* The expansion of $(a + b)^n$, where n is a positive integer, is

(a) $$(a + b)^n = a^n + na^{n-1}b + \frac{n(n-1)}{2!}a^{n-2}b^2 + \frac{n(n-1)(n-2)}{3!}a^{n-3}b^3 + \cdots + \frac{n(n-1)(n-2)\cdots(n-r+2)}{(r-1)!}a^{n-r+1}b^{r-1} + \cdots + b^n;$$

(b) The rth term in this expansion is

$$\frac{n(n-1)(n-2)\cdots(n-r+2)}{(r-1)!}a^{n-r+1}b^{r-1}.$$

(4) *Logarithms.*

(a) If $a^b = x$, then, by definition of logarithm, $\log_a x = b$,

(b) $\log_b a = \dfrac{1}{\log_a b}$,

To any base:

(c) $\log MN = \log M + \log N$,

(d) $\log M^A = A \log M$,

(e) $\log \dfrac{M}{N} = \log MN^{-1} = \log M - \log N$,

(f) $\log \sqrt[n]{M} = \log M^{\frac{1}{n}} = \dfrac{1}{n}\log M$.

(5) *Determinants.* The left-hand member of the identity

(a) $$\begin{vmatrix} a_1 & b_1 \\ a_2 & b_2 \end{vmatrix} = a_1b_2 - a_2b_1$$

is called a determinant of the second order. It is another, and useful, way of writing the algebraic quantity on the right. Similarly for a determinant of the third order:

(b) $$\begin{vmatrix} a_1 & b_1 & c_1 \\ a_2 & b_2 & c_2 \\ a_3 & b_3 & c_3 \end{vmatrix} = a_1b_2c_3 + a_2b_3c_1 + a_3b_1c_2 - a_3b_2c_1 - a_2b_1c_3 - a_1b_3c_2.$$

Determinants (a) and (b) are said to be "expanded" into their equivalent algebraic forms.

(6) *Simultaneous equations.*

(a) Two linear equations in two unknowns.

$$a_1x + b_1y = c_1,$$
$$a_2x + b_2y = c_2.$$

The solution is, in determinant form,

$$x = \frac{\begin{vmatrix} c_1 & b_1 \\ c_2 & b_2 \end{vmatrix}}{D}, \quad y = \frac{\begin{vmatrix} a_1 & c_1 \\ a_2 & c_2 \end{vmatrix}}{D}, \quad D = \begin{vmatrix} a_1 & b_1 \\ a_2 & b_2 \end{vmatrix} \neq 0.$$

(b) Three linear equations in three unknowns.

$$\begin{aligned} a_1x + b_1y + c_1z &= d_1, \\ a_2x + b_2y + c_2z &= d_2, \\ a_3x + b_3y + c_3z &= d_3, \end{aligned} \qquad D = \begin{vmatrix} a_1 & b_1 & c_1 \\ a_2 & b_2 & c_2 \\ a_3 & b_3 & c_3 \end{vmatrix} \neq 0,$$

$$x = \frac{\begin{vmatrix} d_1 & b_1 & c_1 \\ d_2 & b_2 & c_2 \\ d_3 & b_3 & c_3 \end{vmatrix}}{D}, \quad y = \frac{\begin{vmatrix} a_1 & d_1 & c_1 \\ a_2 & d_2 & c_2 \\ a_3 & d_3 & c_3 \end{vmatrix}}{D}, \quad z = \frac{\begin{vmatrix} a_1 & b_1 & d_1 \\ a_2 & b_2 & d_2 \\ a_3 & b_3 & d_3 \end{vmatrix}}{D}.$$

(c) One linear and one quadratic equation, each in two unknowns.

$$\begin{aligned} ax + by + c &= 0, \\ Ax^2 + Bxy + Cy^2 + Dx + Ey + F &= 0. \end{aligned}$$

Solve the linear equation for one of the variables, say x, in terms of the other and substitute this value for x into the quadratic equation. This will result in one quadratic equation in one letter (y) alone which can be solved by the quadratic equation formula, yielding two possible values of y. Substituting these values back into the linear equation will give two corresponding values of x.

(7) *Three special relations.* There are three special ratios that the student should be thoroughly familiar with. These are $0/A$, $A/0$ $(A \neq 0)$, and $0/0$.

Let

$0/A = x$, i.e. $0 = Ax$. There is only one value of x that will yield zero when multiplied by A, namely zero itself. Hence

(a) $$\frac{0}{A} = 0.$$

Let

$A/0 = x$, i.e. $A = 0x$. There is no value of x which, when multiplied by zero, will yield A. Division by zero is impossible or yields infinity. Hence

(b) $$\frac{A}{0} = \infty.$$

Let $0/0 = x$, i.e. $0 = 0x$. Any number x will satisfy this equation. Hence

(c) $$\frac{0}{0} \text{ is indeterminate.}$$

3. Geometry.

(1) *Radian measure.*

(a) A central angle, subtended by an arc equal in length to the radius of the circle, is called a *radian* (Fig. 1).

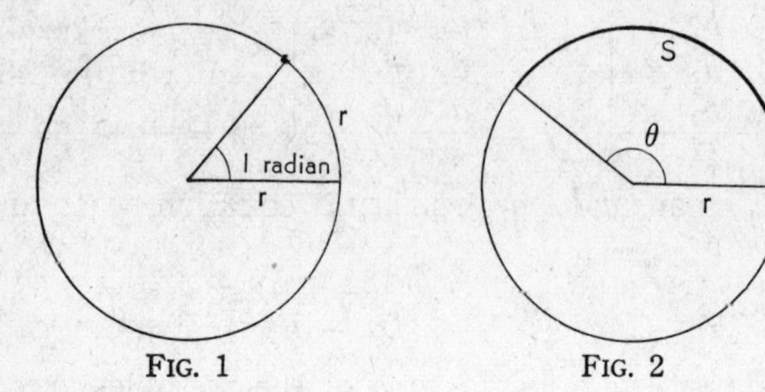

FIG. 1 FIG. 2

(b) If r is the radius of a circle and if θ, measured in radians, is the central angle subtended by an arc S, then (Fig. 2)

$$S = r\theta.$$

(c) Relation between degree measure and radian measure:

$$360^\circ = 2\pi \text{ radians} = 1 \text{ revolution (or circumference).}$$

(2) *Mensuration formulae.* Let r denote radius; θ, central angle in radians; S, arc; h, altitude; b, length of base; s, slant height; A, area of base.

	CIRCUMFERENCE	AREA	VOLUME
(a) Circle	$2\pi r$	πr^2	
(b) Circular sector	$S = r\theta$	$\frac{1}{2} r^2\theta$	
(c) Triangle		$\frac{1}{2} bh$	
(d) Trapezoid		$\frac{1}{2}(b_1 + b_2)h$	
(e) Prism			Ah
(f) Right circular cylinder (limiting case of a prism)		$2\pi rh$	$Ah = \pi r^2 h$
(g) Pyramid			$\frac{1}{3} Ah$
(h) Right circular cone (limiting case of a pyramid)		$\pi rs = \pi r\sqrt{r^2 + h^2}$	$\frac{1}{3}\pi r^2 h$
(i) Sphere		$4\pi r^2$	$\frac{4}{3}\pi r^3$

(j) Similar areas are to each other as the squares of corresponding dimensions.

(k) Similar volumes are to each other as the cubes of corresponding dimensions.

(3) *Pythagorean theorem.* The square on the hypotenuse of any right triangle is equal to the sum of the squares on the two sides: $c^2 = a^2 + b^2$ (Fig. 3).

(4) *Definitions.*

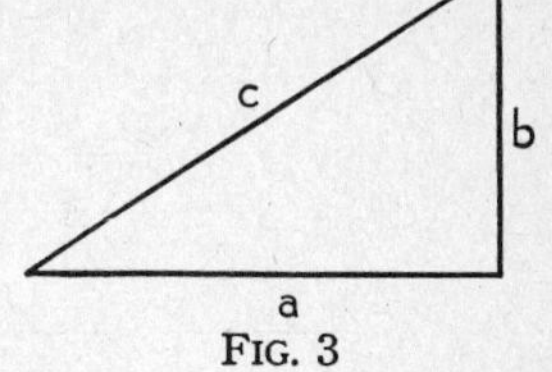

FIG. 3

(a) A median of a triangle is a line joining the midpoint of a side and the opposite vertex.

(b) A rhombus is an equilateral parallelogram.

(c) An isosceles trapezoid is a trapezoid in which the nonparallel sides are equal in length and make equal interior angles with the base.

(d) A convex polygon is a polygon each interior angle of which is less than 180°.

4. Trigonometry.

(1) *Definitions.*

(a) $\sin x = \dfrac{\text{ordinate}}{\text{distance}}$, (d) $\csc x = \dfrac{\text{dist}}{\text{ord}}$,

(b) $\cos x = \dfrac{\text{abscissa}}{\text{distance}}$, (e) $\sec x = \dfrac{\text{dist}}{\text{abs}}$,

(c) $\tan x = \dfrac{\text{ord}}{\text{abs}}$, (f) $\cot x = \dfrac{\text{abs}}{\text{ord}}$.

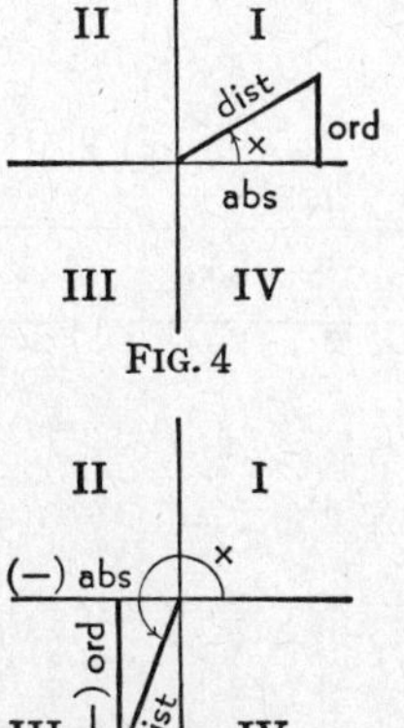

FIG. 4

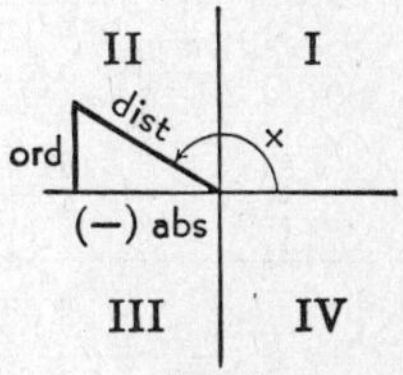

FIG. 5

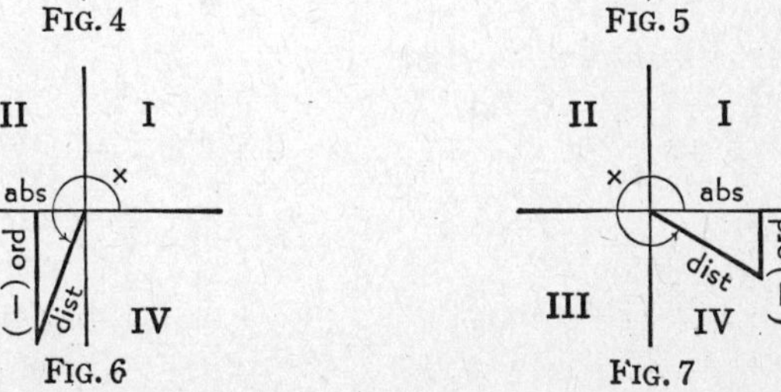

FIG. 6

FIG. 7

(2) *Signs of the trigonometric functions.*

QUADRANT	SIN	COS	TAN	CSC	SEC	COT
I	+	+	+	+	+	+
II	+	−	−	+	−	−
III	−	−	+	−	−	+
IV	−	+	−	−	+	−

(3) *Functions of special angles.*

QUADRANT	DEGREES	RADIANS	SIN	COS	TAN	CSC	SEC	COT
I	0, 360	$0, 2\pi$	0	1	0	∞	1	∞
	30	$\frac{1}{6}\pi$	$\frac{1}{2}$	$\frac{1}{2}\sqrt{3}$	$\frac{1}{3}\sqrt{3}$	2	$\frac{2}{3}\sqrt{3}$	$\sqrt{3}$
	45	$\frac{1}{4}\pi$	$\frac{1}{2}\sqrt{2}$	$\frac{1}{2}\sqrt{2}$	1	$\sqrt{2}$	$\sqrt{2}$	1
	60	$\frac{1}{3}\pi$	$\frac{1}{2}\sqrt{3}$	$\frac{1}{2}$	$\sqrt{3}$	$\frac{2}{3}\sqrt{3}$	2	$\frac{1}{3}\sqrt{3}$
II	90	$\frac{1}{2}\pi$	1	0	∞	1	∞	0
	120	$\frac{2}{3}\pi$	$\frac{1}{2}\sqrt{3}$	$-\frac{1}{2}$	$-\sqrt{3}$	$\frac{2}{3}\sqrt{3}$	-2	$-\frac{1}{3}\sqrt{3}$
	135	$\frac{3}{4}\pi$	$\frac{1}{2}\sqrt{2}$	$-\frac{1}{2}\sqrt{2}$	-1	$\sqrt{2}$	$-\sqrt{2}$	-1
	150	$\frac{5}{6}\pi$	$\frac{1}{2}$	$-\frac{1}{2}\sqrt{3}$	$-\frac{1}{3}\sqrt{3}$	2	$-\frac{2}{3}\sqrt{3}$	$-\sqrt{3}$
III	180	π	0	-1	0	∞	-1	∞
	210	$\frac{7}{6}\pi$	$-\frac{1}{2}$	$-\frac{1}{2}\sqrt{3}$	$\frac{1}{3}\sqrt{3}$	-2	$-\frac{2}{3}\sqrt{3}$	$\sqrt{3}$
	225	$\frac{5}{4}\pi$	$-\frac{1}{2}\sqrt{2}$	$-\frac{1}{2}\sqrt{2}$	1	$-\sqrt{2}$	$-\sqrt{2}$	1
	240	$\frac{4}{3}\pi$	$-\frac{1}{2}\sqrt{3}$	$-\frac{1}{2}$	$\sqrt{3}$	$-\frac{2}{3}\sqrt{3}$	-2	$\frac{1}{3}\sqrt{3}$
IV	270	$\frac{3}{2}\pi$	-1	0	∞	-1	∞	0
	300	$\frac{5}{3}\pi$	$-\frac{1}{2}\sqrt{3}$	$\frac{1}{2}$	$-\sqrt{3}$	$-\frac{2}{3}\sqrt{3}$	2	$-\frac{1}{3}\sqrt{3}$
	315	$\frac{7}{4}\pi$	$-\frac{1}{2}\sqrt{2}$	$\frac{1}{2}\sqrt{2}$	-1	$-\sqrt{2}$	$\sqrt{2}$	-1
	330	$\frac{11}{6}\pi$	$-\frac{1}{2}$	$\frac{1}{2}\sqrt{3}$	$-\frac{1}{3}\sqrt{3}$	-2	$\frac{2}{3}\sqrt{3}$	$-\sqrt{3}$

$\sqrt{2} = 1.414$, $\frac{1}{2}\sqrt{2} = .707$; $\sqrt{3} = 1.732$, $\frac{1}{2}\sqrt{3} = .866$, $\frac{1}{3}\sqrt{3} = .577$

(4) *Fundamental identities.*

(a) $\sin x = \dfrac{1}{\csc x}$, (b) $\cos x = \dfrac{1}{\sec x}$,

(c) $\tan x = \dfrac{1}{\cot x}$, (d) $\tan x = \dfrac{\sin x}{\cos x}$,

(e) $\sin^2 x + \cos^2 x = 1$, (f) $1 + \tan^2 x = \sec^2 x$,

(g) $1 + \cot^2 x = \csc^2 x$.

(5) *Reduction formulae rule:*

1st. Any trigonometric function of the angle $\left(k\frac{\pi}{2} \pm \alpha\right)$ is equal to $(\pm)$ the same function of α, if k is even, and is equal to $(\pm)$ the cofunction of α if k is odd.

2nd. The "+" sign is used if the original function of the original angle $\left(k\frac{\pi}{2} \pm \alpha\right)$ is plus; the "−" sign is used if the original function is negative. The sign of the original function of $\left(k\frac{\pi}{2} \pm \alpha\right)$ is determined by the usual quadrantal conventions. To summarize:

$$\text{Any function of}\left(k\frac{\pi}{2} \pm \alpha\right) = \pm\begin{cases}\text{Same function of } \alpha, \text{ if } k \text{ is even;} \\ \text{cofunction of } \alpha, \text{ if } k \text{ is odd.} \\ \text{Use sign of original function of} \\ \left(k\frac{\pi}{2} \pm \alpha\right).\end{cases}$$

(6) *Functions of the sum and difference of two angles.*

(a) $\sin(x \pm y) = \sin x \cos y \pm \cos x \sin y$,

(b) $\cos(x \pm y) = \cos x \cos y \mp \sin x \sin y$,

(c) $\tan(x \pm y) = \dfrac{\tan x \pm \tan y}{1 \mp \tan x \tan y}$.

(7) *Multiple angle formulae.*

(a) $\sin 2x = 2 \sin x \cos x$,

(b) $\cos 2x = \cos^2 x - \sin^2 x = 2\cos^2 x - 1 = 1 - 2\sin^2 x$,

(c) $\tan 2x = \dfrac{2 \tan x}{1 - \tan^2 x}$,

(d) $\sin \dfrac{x}{2} = \sqrt{\dfrac{1 - \cos x}{2}}$,

(e) $\cos \dfrac{x}{2} = \sqrt{\dfrac{1 + \cos x}{2}}$,

(f) $\tan \dfrac{x}{2} = \sqrt{\dfrac{1 - \cos x}{1 + \cos x}} = \dfrac{1 - \cos x}{\sin x} = \dfrac{\sin x}{1 + \cos x}$.

(8) *Sum and product formulae.*

(a) $\sin x + \sin y = 2 \sin \frac{1}{2}(x + y) \cos \frac{1}{2}(x - y)$,

(b) $\sin x - \sin y = 2 \cos \frac{1}{2}(x + y) \sin \frac{1}{2}(x - y)$,

(c) $\cos x + \cos y = 2 \cos \frac{1}{2}(x + y) \cos \frac{1}{2}(x - y)$,
(d) $\cos x - \cos y = -2 \sin \frac{1}{2}(x + y) \sin \frac{1}{2}(x - y)$,
(e) $\sin x \sin y = \frac{1}{2} \cos (x - y) - \frac{1}{2} \cos (x + y)$,
(f) $\sin x \cos y = \frac{1}{2} \sin (x - y) + \frac{1}{2} \sin (x + y)$,
(g) $\cos x \cos y = \frac{1}{2} \cos (x - y) + \frac{1}{2} \cos (x + y)$.

(9) *Formulae for plane triangles.* Let a, b, c be sides; A, B, C, opposite angles; $s = \dfrac{a + b + c}{2}$, semi-perimeter;

$$r = \sqrt{\frac{(s - a)(s - b)(s - c)}{s}},$$

radius of the inscribed circle; R, radius of the circumscribed circle; K, area.

(a) Law of sines: $\dfrac{a}{\sin A} = \dfrac{b}{\sin B} = \dfrac{c}{\sin C} = 2R$,

(b) Law of cosines: $a^2 = b^2 + c^2 - 2bc \cos A$,

(c) Law of tangents: $\dfrac{a + b}{a - b} = \dfrac{\tan \frac{1}{2}(A + B)}{\tan \frac{1}{2}(A - B)}$,

(d) Tangent of half angle: $\tan \frac{1}{2} A = \dfrac{r}{s - a}$,

(e) Area:

$$K = \tfrac{1}{2} ab \sin C = \sqrt{s(s - a)(s - b)(s - c)} = rs = \frac{abc}{4R}.$$

PLANE ANALYTIC GEOMETRY

CHAPTER II

FUNDAMENTAL CONCEPTS

5. Introduction. Analytic geometry, or the analytic treatment of geometry, was introduced by René Descartes in his *La Géométrie* published in 1637. Accordingly, after the name of its founder, analytic or coordinate geometry is often referred to as *Cartesian* geometry. It is essentially a method of studying geometry by means of algebra. Earlier mathematicians had continued to resort to the conventional methods of geometric reasoning as set forth in great detail by Euclid and his school some 2,000 years before. The tremendous advances made in the study of geometry since the time of Descartes are largely due to his introduction of the coordinate system and the associated algebraic or analytic methods. And, conversely, the use of analytic geometry in the study of equations has been of direct benefit to algebra.

6. Rectangular Coordinates. Consider two perpendicular lines $X'X$ and $Y'Y$ intersecting in the point O (Fig. 8). $X'X$

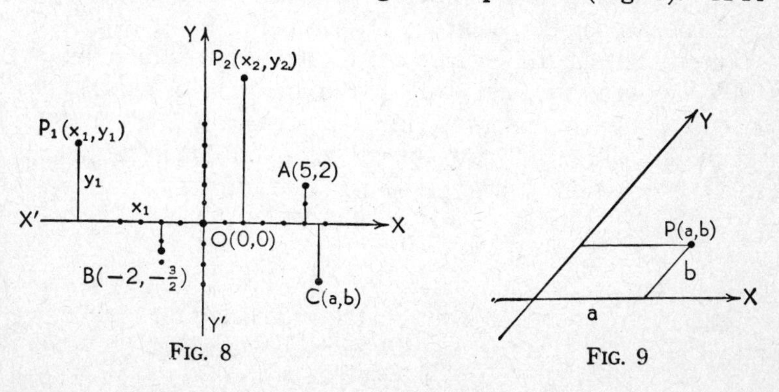

FIG. 8

FIG. 9

is called the *axis of X*, $Y'Y$ the *axis of Y*, and together they form a *rectangular coordinate system.* The axes divide the plane into four quadrants which are usually labeled as in trigonometry. The point O is called the *origin.* When numerical scales are established on the axes, positive distances x (*abscissae*) are laid off to the right of the origin, negative abscissae to the left; positive distances y (*ordinates*) are drawn upwards and negative ordinates downward. Thus OX, OY have positive sense (or direction) while OX', OY' have negative sense.

Clearly such a system of coordinates can be used to describe the position of points in the plane. By going out $+5$ units on the X-axis and $+2$ units on the Y-axis, for example, the point A is located. The point A is said to have the pair of numbers 5 and 2 as its *coordinates*, and it is customary to write $A(5, 2)$ or simply $(5, 2)$. Similarly B has the coordinates $(-2, -\frac{3}{2})$ and lies in the third quadrant. It is evident that for the point P_1, pictured in the second quadrant, the x-coordinate is negative and the y-coordinate is positive. We still may write $P_1(x_1, y_1)$, letting x_1 itself include the minus sign. P_2 is a point in the first quadrant with x_2 and y_2 both positive; for the point C the abscissa a is positive, the ordinate b is negative. The coordinates of the origin are $(0, 0)$.

The *fundamental principle* here is that there is a one-to-one correspondence between number pairs and points in the plane: to each pair of numbers there corresponds one and only one point and, conversely, to each point in the plane there corresponds one and only one pair of numbers.

A Cartesian coordinate system may also be established by means of two non-perpendicular lines as in Fig. 9. The point P has coordinates a and b where a is measured in the direction of the X-axis and b is measured in the direction of the Y-axis. Such a non-rectangular system is occasionally very useful, but in general it leads to technical complications in formulae. (See the equation of a hyperbola referred to its asymptotes, p. 96.)

7. Distance between Two Points. Let $P_1(x_1, y_1)$ and $P_2(x_2, y_2)$ be two points lying in the first quadrant and draw P_1Q, QP_2 parallel to the coordinate axes. (See Fig. 10.) By simple subtraction of abscissae, $P_1Q = x_2 - x_1$; similarly, subtracting ordinates, $QP_2 = y_2 - y_1$. Making use of the square on the hypotenuse of right triangle P_1QP_2, we have (see Pythagorean theorem, p. 5)

$$\overline{P_1P_2}^2 = (x_2 - x_1)^2 + (y_2 - y_1)^2 \tag{1}$$

and the positive distance P_1P_2 (call it d) is given by

$$d = \sqrt{(x_2 - x_1)^2 + (y_2 - y_1)^2}. \tag{2}$$

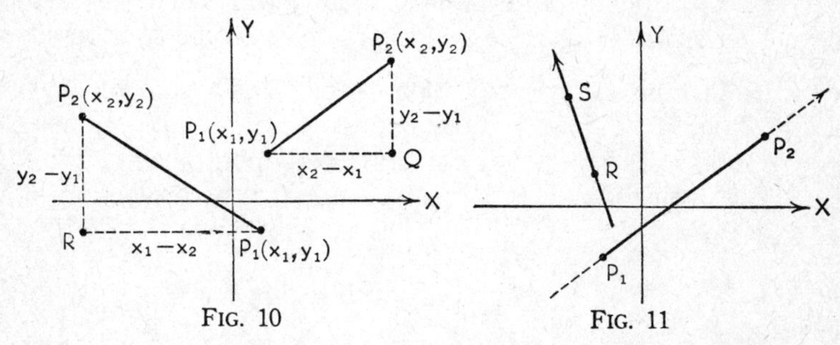

FIG. 10 FIG. 11

The same formula holds true regardless of the quadrants in which the points lie and regardless of the order in which the points are taken. For example, the positive distance $RP_1 = x_1 - x_2 = -(x_2 - x_1)$ and the positive distance $RP_2 = y_2 - y_1$ since these distances are measured in the positive directions of the axes. But $(x_1 - x_2)^2 = (x_2 - x_1)^2$ and (2) holds as before.

Illustration. Find the distance between the two points $(3, -1)$, $(-4, -2)$.

Solution. Taking the points in the given order, we have

$$\begin{aligned} d &= \sqrt{(-4 - 3)^2 + (-2 + 1)^2} \\ &= \sqrt{50} = 5\sqrt{2}. \end{aligned}$$

Or again, reversing this order,

$$\begin{aligned} d &= \sqrt{(3 + 4)^2 + (-1 + 2)^2} \\ &= 5\sqrt{2}. \end{aligned}$$

8. Directed Line Segments. Although the distance between two points is usually considered positive, yet it is at times desirable to associate with a line segment *direction* or *sense*. This amounts to attaching a plus or minus sign to the segment according to some convention. But since there is little agreement among authors on this point we shall assume that if sense is important in a particular case it will be specified by the order in which the end points are given. Thus, in Fig. 11, if P_1P_2 is the given positively directed segment then P_2P_1 has the negative sense and we write $P_1P_2 = -P_2P_1$. The direction of a line, if essential to the immediate argument, will be denoted by the order in which two points on it are specified. Thus if *RS* is the positive direction *SR* will be the negative.

The context of the material under study will usually make it clear whether both distance and direction are to be considered.

9. Projections. The projection of the signed segment A_1A_2 (Fig. 12) upon the line L is, by definition, the signed segment P_1P_2 cut off by the perpendiculars dropped upon L from A_1 and A_2. The projection of A_2A_1 upon L is P_2P_1; the projection of A_2A_3 is P_2P_3.

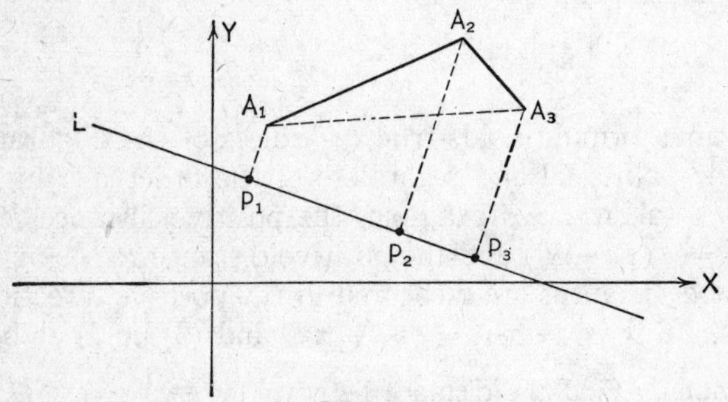

FIG. 12

As in mechanics we may think of A_1A_2, A_2A_3 as being free vectors, the vector sum of which is A_1A_3. With this definition of vector sum in mind we note that the *sum of the projections* of A_1A_2 and A_2A_3 upon L is equal to the *projection of the sum* A_1A_3 upon L. This is a general property of projections regardless of the number and orientation of the segments. If the individual segments, or vectors, represent forces, then the vector sum represents the *resultant* force, and if the vectors form a closed polygon the resultant is zero and a state of equilibrium exists.

The projections of a line segment upon the axes, or upon lines parallel to them, constitute important special cases. For example, in Fig. 13 below, PS, or QR, is the projection of PP_2 in the x direction while PQ, or SR, is the projection of PP_1 in the y direction.

10. Point of Division. Given a directed line segment such as P_1P_2 in Fig. 13; to find the coordinates of the point P which divides P_1P_2 in a given ratio r_1/r_2. Let P have the coordinates (x, y) which are to be determined. Sense is important here and P must be located so that $P_1P/PP_2 = r_1/r_2$.

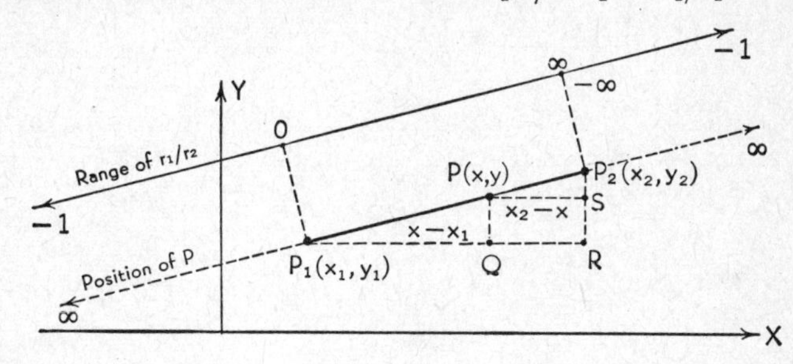

FIG. 13

Now, by similar triangles, $(x - x_1)/r_1 = (x_2 - x)/r_2$, from which it follows that

(1) $$x = \frac{x_1r_2 + x_2r_1}{r_1 + r_2}.$$

Similarly

(2) $$y = \frac{y_1r_2 + y_2r_1}{r_1 + r_2}.$$

For the midpoint of the segment P_1P_2 the ratio r_1/r_2 must be unity; hence $r_1 = r_2$ and (1) and (2) specialize to

(3) $$\bar{x} = \frac{x_1 + x_2}{2}, \quad \bar{y} = \frac{y_1 + y_2}{2}.$$

Formulae (1), (2), and (3) have useful physical interpretations. In (1) and (2) x and y are the coordinates of the center of gravity of masses r_2 and r_1 placed respectively at P_1 and P_2. If the masses are equal, the center of gravity lies halfway between them as indicated by (3).

It is of further interest to note the positions of P for various values of the ratio r_1/r_2. If this ratio is zero, then P coincides with P_1, and if this ratio is a positive number, P is an internal point of division. As $r_1/r_2 \to +\infty$, $P \to P_2$. For $-\infty < r_1/r_2 < -1$, P is an external point of division (in the direction P_1P_2) with P_1P positive and PP_2 negative. For $-1 < r_1/r_2 < 0$, P is an external point in the opposite direction with P_1P negative and PP_2 positive.

Illustration 1. Find the coordinates of the midpoint of the segment $P_1(3, 7)$, $P_2(-2, 3)$.

Solution. $\bar{x} = \dfrac{3 + (-2)}{2} = \dfrac{1}{2}$, $\bar{y} = \dfrac{7 + 3}{2} = 5$.

Illustration 2. Find the coordinates of the point P which divides the segment $P_1(-2, 5)$, $P_2(4, -1)$ in the ratio of (a) $r_1/r_2 = \frac{6}{5}$; (b) $r_1/r_2 = -2$; (c) $r_1/r_2 = -\frac{1}{3}$.

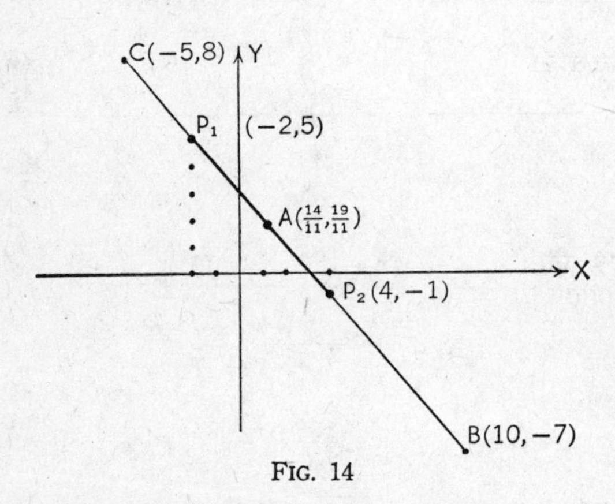

FIG. 14

Solution. (a) $x = \dfrac{(-2)(5) + (4)(6)}{11} = \dfrac{14}{11}$,

$$y = \frac{(5)(5) + (-1)(6)}{11} = \frac{19}{11};$$

(b) $x = \dfrac{(-2)(1) + (4)(-2)}{-1} = 10$,

$$y = \frac{(5)(1) + (-1)(-2)}{-1} = -7;$$

(c) $x = \dfrac{(-2)(3) + (4)(-1)}{2} = -5$,

$$y = \frac{(5)(3) + (-1)(-1)}{2} = 8.$$

Note that, in terms of positive distances, $P_1P_2 = P_2B = 6\sqrt{2}$ and $CP_1 = \frac{1}{2} P_1P_2 = 3\sqrt{2}$. These are in agreement with the values of the ratios given in (b) and (c).

11. Inclination, Slope, Direction Cosines. The angle $\theta(0 \leq \theta < 180°)$, measured counterclockwise from the positive X-axis to a line, directed or not, is called the *inclination* of the line. The tangent of this angle, $\tan \theta$, and generally designated by the letter m, is called the *slope* of the line. It is evident from Fig. 15 that the slope is given by

$$\text{(1)} \qquad \text{Slope of } P_1P_2 = \tan \theta = m = \frac{y_2 - y_1}{x_2 - x_1}.$$

This formula is independent of the position and order of the two points involved.

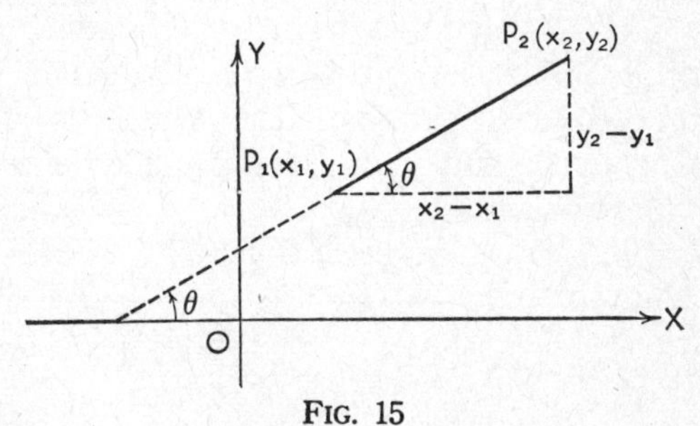

FIG. 15

Although $\tan \theta$ is a most natural trigonometric function to use in describing the general trend, or steepness, of a line, it is not the only one that can be used. Indeed, we need not restrict ourselves to the one angle θ. Let the given line have an established sense P_1P_2 (Fig. 16) and call by α and β respectively the two *direction angles* made by the positive direction of the line and the positive directions of the axes. By definition $\lambda = \cos \alpha$ and $\mu = \cos \beta$ are called the *direction cosines* of the line. Any two numbers proportional to the direction cosines are called *direction numbers* of the line. Thus $a = k\lambda = k \cos \alpha$ and $b = k\mu = k \cos \beta$ are direction numbers. A sensed line has only one set of direction cosines since α and β are then unique. If the sense of a line is reversed the angles α and β are replaced by their supplements, so that a line without sense has two sets

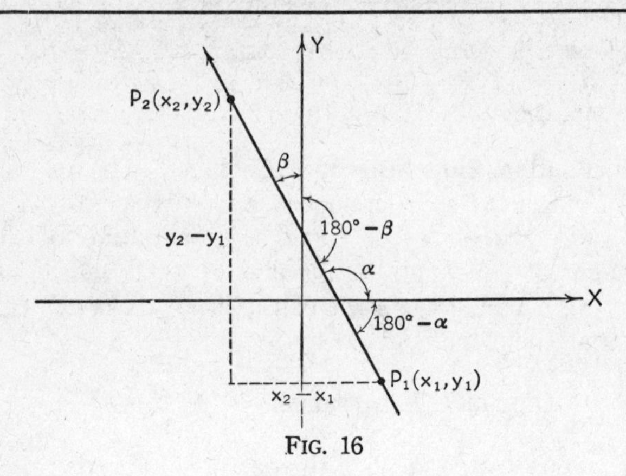

FIG. 16

of direction cosines: $\cos \alpha$, $\cos \beta$ and $\cos (180° - \alpha)$, $\cos (180° - \beta)$, i.e., λ, μ and $-\lambda$, $-\mu$, corresponding to the two directions of the line. In this case we call either λ, μ or $-\lambda$, $-\mu$ the direction cosines. Note that either λ or μ (or both) can be negative numbers depending on the sense and general trend of the line.

Directly from the figure we see that

$$\cos \alpha = \frac{x_2 - x_1}{d}, \quad \cos \beta = \frac{y_2 - y_1}{d}, \tag{2}$$

whence

$$x_2 - x_1 = d \cos \alpha, \quad y_2 - y_1 = d \cos \beta. \tag{3}$$

Thus the differences in the respective coordinates of any two points on a line, namely $x_2 - x_1$ and $y_2 - y_1$, are direction numbers of the line, the proportionality constant being the distance between the two points.

Upon squaring and adding equations (2), remembering that $d^2 = (x_2 - x_1)^2 + (y_2 - y_1)^2$, we obtain

$$\cos^2 \alpha + \cos^2 \beta = 1. \tag{4}$$

This can also be written $\lambda^2 + \mu^2 = 1$ and is an important identity connecting the direction cosines of a line. The direction cosines of a line parallel to the X-axis are $\lambda = \pm 1$, $\mu = 0$; the direction ccsines of a line parallel to the Y-axis are $\lambda = 0$, $\mu = \pm 1$. No distinction is made between a line and a line segment.

If a and b are direction numbers of a line, $a = k\lambda$ and $b = k\mu$.

Since $\lambda^2 + \mu^2 = 1$, it follows that $a^2/k^2 + b^2/k^2 = 1$ or $k^2 = a^2 + b^2$. Thus the direction cosines themselves are

$$(5) \qquad \lambda = \frac{a}{\sqrt{a^2 + b^2}}, \quad \mu = \frac{b}{\sqrt{a^2 + b^2}}.$$

Throughout plane and solid analytic geometry much use is made of the notions of slope and direction cosines and the student should make every attempt to master these ideas before proceeding. He should also be thoroughly aware of any differences that may exist between the basic assumptions made by the author of this Outline and those of the particular author he is following.

Illustration 1. Find the slope of the line joining the points $P_1(1, 2)$ and $P_2(-5, 3)$.

Solution. $$m = \frac{y_2 - y_1}{x_2 - x_1} = \frac{3 - 2}{-5 - 1} = -\frac{1}{6}.$$

Illustration 2. Find the direction cosines of the sensed line $P_1(1, 2)$, $P_2(-5, 3)$.

Solution. $$\lambda = \cos \alpha = \frac{x_2 - x_1}{d} = \frac{-5 - 1}{\sqrt{37}} = -\frac{6}{\sqrt{37}},$$

$$\mu = \cos \beta = \frac{y_2 - y_1}{d} = \frac{3 - 2}{\sqrt{37}} = \frac{1}{\sqrt{37}}.$$

Illustration 3. Direction numbers of a given line are -2 and 3. Find the direction cosines.

Solution. $$\lambda = \frac{a}{\sqrt{a^2 + b^2}} = \frac{-2}{\sqrt{(-2)^2 + 3^2}} = -\frac{2}{\sqrt{13}},$$

$$\mu = \frac{b}{\sqrt{a^2 + b^2}} = \frac{3}{\sqrt{13}}.$$

These may also be taken as $\lambda = +2/\sqrt{13}$ and $\mu = -3/\sqrt{13}$.

12. Parallel and Perpendicular Lines. If two lines are parallel they have the same slope, and if in addition they have the same sense they have the same direction cosines. Sometimes lines which are parallel but which have opposite sense are called *antiparallel;* antiparallel lines have the same slope but the direction cosines of the one are the negative values of the direction cosines of the other.

If two lines L_1 and L_2 are perpendicular the slope of one is the negative reciprocal of the other. For, from Fig. 17,

$$m_1 = \tan \theta_1 = \tan (90° + \theta_2) = -\cot \theta_2,$$
$$= \frac{1}{-\tan \theta_2}.$$

Hence

(1) $$m_1 = -\frac{1}{m_2}.$$

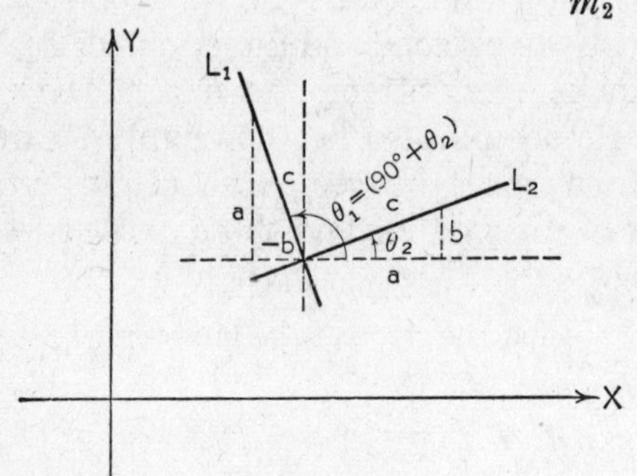

FIG. 17

The student should check the trigonometry by making use of the reduction rule (5), Chapter I.

The direction cosines λ_1, μ_1 and λ_2, μ_2 of two perpendicular lines are connected by the relations

(2) $$\lambda_1 = \pm\mu_2, \quad \mu_1 = \mp\lambda_2.$$

Illustration 1. Find the slope of a line which is perpendicular to the line joining $P_1(2, 4)$, $P_2(-2, 1)$.

Solution. The slope of the line P_1P_2 is

$$m_1 = \frac{1-4}{-2-2} = \frac{3}{4}.$$

Therefore the slope of a perpendicular line is

$$m_2 = -\frac{4}{3}.$$

Illustration 2. Find the direction cosines of a line which is perpendicular to the sensed line $P_1(2, 4)$, $P_2(-2, 1)$.

Solution. The direction cosines of the directed line P_1P_2 are

$$\lambda_1 = -\tfrac{4}{5}, \quad \mu_1 = -\tfrac{3}{5}.$$

Hence those of a perpendicular line (without sense) are

$$\lambda_2 = \pm\tfrac{3}{5}, \quad \mu_2 = \mp\tfrac{4}{5}.$$

13. Angle between Two Lines. The angle between two nonintersecting lines is 0° or 180° according as the lines are parallel or antiparallel. In the case of intersecting lines we do not calculate the angle directly but compute instead some trigonometric function of the angle, generally tangent or cosine.

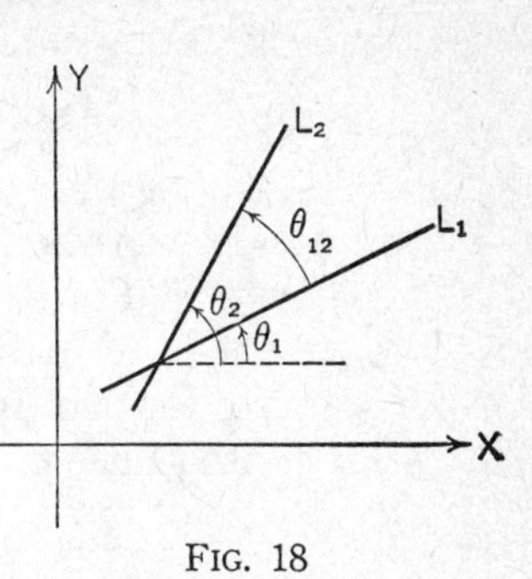

FIG. 18

I. *Angle in terms of slopes.* Let θ_{12} be the angle between two intersecting lines measured counterclockwise *from* line L_1 *to* line L_2 (Fig. 18). Since $\theta_{12} = \theta_2 - \theta_1$ it follows that

$$\tan \theta_{12} = \tan (\theta_2 - \theta_1) = \frac{\tan \theta_2 - \tan \theta_1}{1 + \tan \theta_2 \tan \theta_1},$$

which, in terms of the slopes of the lines, yields

$$\text{(1)} \qquad \tan \theta_{12} = \frac{m_2 - m_1}{1 + m_2 m_1}.$$

The subscripts on θ_{12} are put there to emphasize that this is the particular angle as defined: the angle *from* L_1 *to* L_2 measured counterclockwise. Since the lines may be designated in either order, either one of the two angles that the two lines make with each other can be chosen without guesswork. [The tangent of the supplementary angle $(180° - \theta_{12})$ is given by $\tan (180° - \theta_{12}) = \frac{m_1 - m_2}{1 + m_1 m_2}$.]

II. *Angle in terms of direction cosines.* Instead of using the tangent function we can use the cosine, but here it is best to consider the angle θ between *specified directions* of the two lines. Consider, then, two directed lines L_1 and L_2 with direction cosines λ_1, μ_1 and λ_2, μ_2 respectively (Fig. 19). We have

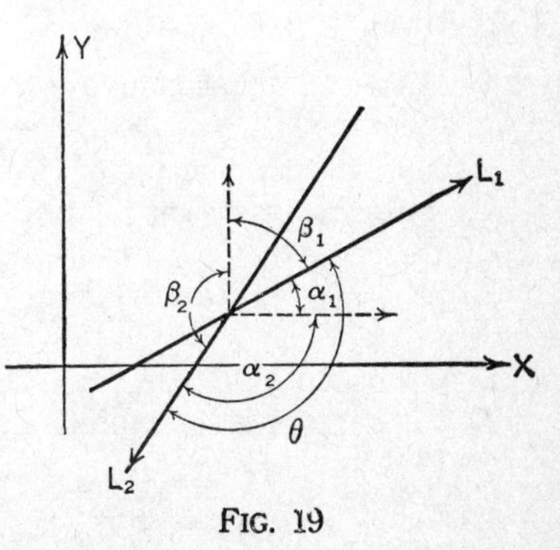

FIG. 19

$$\cos\theta = \cos(\alpha_1 + \alpha_2) = \cos\alpha_1\cos\alpha_2 - \sin\alpha_1\sin\alpha_2.$$

It is left to the student to check, by his knowledge of trigonometry, that $\sin\alpha_1 = \cos\beta_1$ and $\sin\alpha_2 = -\cos\beta_2$. Hence

$$(2) \qquad \cos\theta = \lambda_1\lambda_2 + \mu_1\mu_2.$$

This result is independent of the position and sense of the two lines involved even though the preliminary trigonometry may vary somewhat. For the particular lines directed as indicated in Fig. 19, $\theta = \alpha_1 + \alpha_2$ but this relation is independent neither of sense nor of the arbitrary labels L_1, L_2. The student should draw for himself lines in other positions, noting that in general θ is always one of the forms

$$\theta = \alpha_1 \pm \alpha_2, \quad \theta = \alpha_2 - \alpha_1, \quad \theta = \beta_1 \pm \beta_2, \quad \theta = \beta_2 - \beta_1.$$

He should also satisfy himself that, regardless of the intermediate trigonometry, formula (2) always obtains.

The acute angle between two undirected lines is given by

$$(3) \qquad \cos\theta = |\lambda_1\lambda_2 + \mu_1\mu_2|.$$

If $\theta = 90°$, $\cos\theta = 0$. Hence, from (2), the condition that two lines be perpendicular is that

$$(4) \qquad \lambda_1\lambda_2 + \mu_1\mu_2 = 0$$

or, what is the same thing, that

$$(5) \qquad a_1a_2 + b_1b_2 = 0$$

where a_1, b_1 and a_2, b_2 are direction numbers.

Illustration 1. Find the least angle of the triangle $A(1, 4)$, $B(-5, -1)$, $C(0, -6)$.

Solution. We first compute the slopes of the sides AB, BC, CA. These are $m_{AB} = \frac{5}{6}$, $m_{BC} = -1$, $m_{CA} = 10$.

Next we must make use of (1), taking care to combine the slopes properly so as to get interior angles.

$$\tan BAC = \frac{10 - \frac{5}{6}}{1 + 10(\frac{5}{6})} = \frac{55}{56}.$$

Similarly $\tan ACB = \frac{11}{9}$ and $\tan CBA = 11$.
The angle at A is therefore the smallest.

Illustration 2. Find the angle between the directed lines $P_1(1, 3)$, $P_2(-4, -3)$ and $P_3(2, 0)$, $P_4(-5, 6)$.

Solution. *By slopes.* The angles θ_{12} in question is the counterclockwise angle from P_3P_4 to P_1P_2. Now $m_{P_1P_2} = \frac{6}{5}$ and $m_{P_3P_4} = -\frac{6}{7}$. Hence

$$\tan \theta_{12} = \frac{\frac{6}{5} + \frac{6}{7}}{1 + (\frac{6}{5})(-\frac{6}{7})} = -72.$$

By direction cosines. The direction cosines of P_1P_2 are

$$\lambda_1 = \frac{-5}{\sqrt{61}}, \quad \mu_1 = \frac{-6}{\sqrt{61}}$$

and those of P_3P_4 are

$$\lambda_2 = \frac{-7}{\sqrt{85}}, \quad \mu_2 = \frac{6}{\sqrt{85}}.$$

Therefore

$$\cos \theta = \left(-\frac{5}{\sqrt{61}}\right)\left(-\frac{7}{\sqrt{85}}\right) + \left(-\frac{6}{\sqrt{61}}\right)\left(\frac{6}{\sqrt{85}}\right) = \frac{-1}{\sqrt{61}\sqrt{85}}.$$

Illustration 3. Show that the two lines $P_1(7, 5)$, $P_2(1, 1)$ and $P_3(4, -3)$, $P_4(2, 0)$ are perpendicular.

Solution. *By slopes.* Since the slopes are $\frac{2}{3}$ and $-\frac{3}{2}$ respectively, the lines are perpendicular.

By direction numbers. We may take as direction numbers the differences in the respective coordinates. This yields

$$a_1 = 6, \quad b_1 = 4, \quad a_2 = 2, \quad b_2 = -3;$$

and applying (4), we have

$$a_1a_2 + b_1b_2 = (6)(2) + (4)(-3) = 0.$$

Therefore the lines are perpendicular.

14. Area of a Triangle. Consider the general triangle whose vertices are $A(x_1, y_1)$, $B(x_2, y_2)$, and $C(x_3, y_3)$. Project AC, CB,

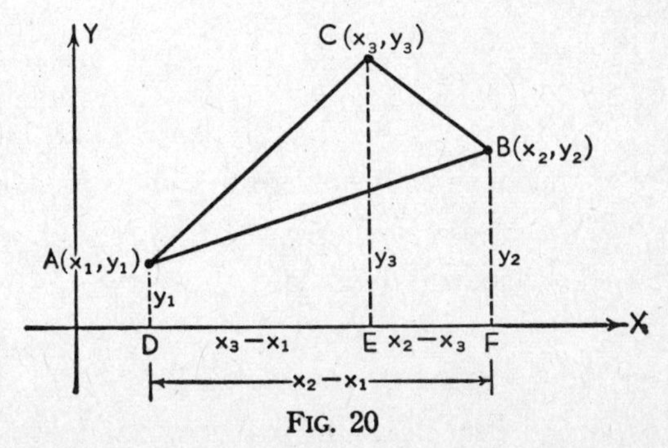

FIG. 20

AB upon the x-axis, obtaining DE, EF, DF respectively. Then (Fig. 20)

(1) Area ABC = Area $ACED$ + Area $CBFE$ − Area $ABFD$.

Since the area of a trapezoid is one-half the sum of the parallel sides multiplied by the distance between them, (1) becomes

$$\text{(2) Area } ABC = \tfrac{1}{2}\{(y_1 + y_3)(x_3 - x_1) + (y_3 + y_2)(x_2 - x_3) - (y_1 + y_2)(x_2 - x_1)\},$$
$$= \tfrac{1}{2}(x_1y_2 + x_2y_3 + x_3y_1 - x_3y_2 - x_2y_1 - x_1y_3)$$

upon simplification.

This can be put into a very compact and easily remembered form by making use of determinants. (See Determinants, p. 2.) The result is

$$\text{(3)} \qquad \text{Area } ABC = \frac{1}{2}\begin{vmatrix} x_1 & y_1 & 1 \\ x_2 & y_2 & 1 \\ x_3 & y_3 & 1 \end{vmatrix}.$$

Caution: The area will be positive by this formula if and only if the vertices are chosen so that, as the perimeter is traversed in the order A, B, C, the area of the triangle lies to the left (the traverse is counterclockwise). If in a given problem the answer turns out to be negative it merely means that the wrong order (clockwise) has been chosen. Attach a plus sign to the final answer in any case.

Illustration 1. Find the area of the triangle (2, 1), (5, −3), (−8, 0).

Solution. Taking the order as given we have

$$\text{Area} = \frac{1}{2}\begin{vmatrix} 2 & 1 & 1 \\ 5 & -3 & 1 \\ -8 & 0 & 1 \end{vmatrix} = \tfrac{1}{2}(-6 + 0 - 8 - 24 - 5 - 0),$$
$$= -\tfrac{43}{2}.$$

The area is $\frac{43}{2}$ square units; the order was chosen clockwise.

Illustration 2. Show that the three points $A(1, 5)$, $B(6, -1)$, $C(-4, 11)$ are collinear (lie on a line).

Solution. *By slopes.*

$$m_{AB} = -\tfrac{6}{5} = m_{BC} = m_{AC}.$$

Since these slopes are the same, the points are collinear.

By distances.

$$AB = \sqrt{25 + 36} = \sqrt{61},$$
$$AC = \sqrt{25 + 36} = \sqrt{61},$$
$$BC = \sqrt{100 + 144} = \sqrt{244} = 2\sqrt{61}.$$

Since $AB + AC = BC$, the points are collinear.

By area.

$$\text{Area } ABC = \tfrac{1}{2}\begin{vmatrix} 1 & 5 & 1 \\ 6 & -1 & 1 \\ -4 & 11 & 1 \end{vmatrix} = \tfrac{1}{2}(-1 + 66 - 20 - 4 - 30 - 11)$$
$$= 0.$$

Since the area of the "triangle" ABC is zero, the points do not form a triangle but instead lie on a straight line.

15. Applications to Elementary Geometry. Of the propositions of plane (and solid) geometry, many are amenable to direct treatment by the methods of analytic geometry. The fundamental properties of a geometric configuration do not in any way depend upon a coordinate system; they depend upon the interrelations of the component parts of the figure. Since this is true, a given problem can often be greatly simplified by the proper choice of axes. Of course care must be exercised in choosing axes so that there will be no loss in generality. For example, if the proposition relates to a general triangle the X-axis can be made to coincide with one side of the triangle with a vertex at the origin. But then there is no further freedom of choice: the Y-axis cannot be passed through the third vertex since otherwise the triangle would not be a general triangle but would be a right triangle instead. (We assume a rectangular coordinate system and not an oblique one, although this is precisely one place where oblique axes are most useful.

Once axes are chosen it becomes necessary to translate the geometric data into coordinates and equations so that the work may proceed algebraically.

Illustration 1. Prove that the diagonals of a rectangle are equal.

Solution. Draw a rectangle $ABCD$ and choose axes as in Fig. 21.

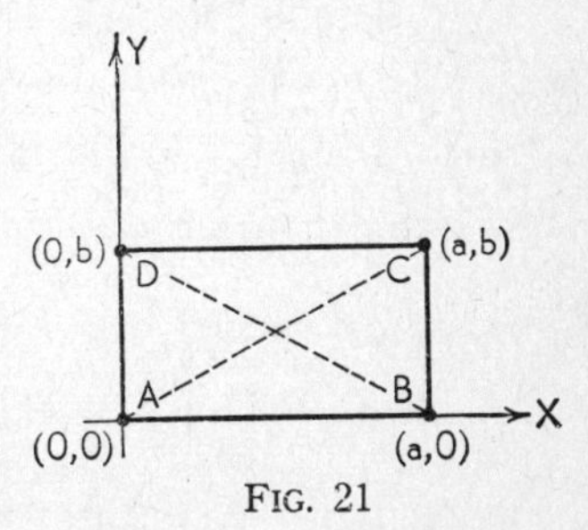

FIG. 21

Then write down general coordinates of the vertices, remembering that the figure is a rectangle. The length of either diagonal then is $\sqrt{a^2 + b^2}$, which proves the proposition.

Illustration 2. Prove that the diagonals of a parallelogram bisect each other.

Solution. Choose axes as in Fig. 22, letting the coordinates of three vertices be $(0, 0)$, (a, b), and $(c, 0)$. Note that the coordinates of the fourth vertex are then fixed; they are $(a + c, b)$.

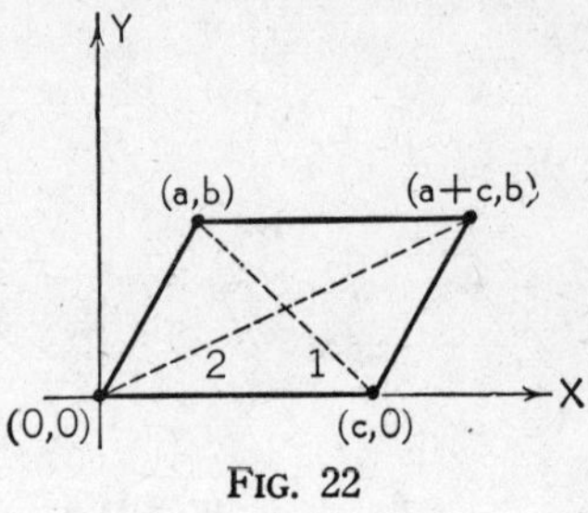

FIG. 22

By formula (3), p. 13, the midpoint of diagonal 1 is $\left(\frac{a+c}{2}, \frac{b}{2}\right)$; the midpoint of diagonal 2 is $\left(\frac{a+c}{2}, \frac{b}{2}\right)$. Since these are the coordinates of the same point, this point lies on each diagonal and the diagonals bisect each other.

Illustration 3. Prove that the medians of a triangle intersect in a point $\frac{2}{3}$ of the distance along one from a vertex toward the opposite side.

Solution. Axes and coordinates are chosen as in Fig. 23;

midpoint of P_1Q_1 is $R_2(a/2, 0)$,

midpoint of Q_1R_1 is $P_2\left(\frac{a+b}{2}, \frac{c}{2}\right)$,

midpoint of R_1P_1 is $Q_2(b/2, c/2)$.

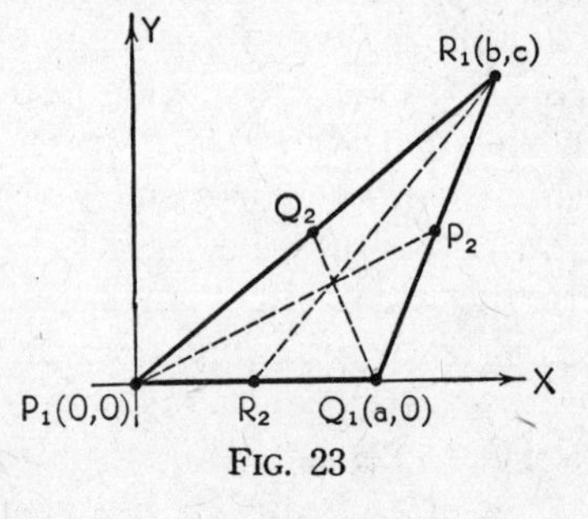

FIG. 23

The point P, $\frac{2}{3}$ of the way from P_1 to P_2, divides P_1P_2 in the ratio of 2 to 1. By formulae (1) and (2), p. 13, the coordinates of P are $P\left(x = \frac{a+b}{3}, y = \frac{c}{3}\right)$. Similarly, the coordinates of the point Q, which is $\frac{2}{3}$ of the way from Q_1 to Q_2, are $Q\left(\frac{a+b}{3}, \frac{c}{3}\right)$ and those of R, which is $\frac{2}{3}$ of the way from R_1 to R_2, are $R\left(\frac{a+b}{3}, \frac{c}{3}\right)$. Since P, Q, R all have the same coordinates, they coincide and the proof of the proposition is complete. (The intersection of the medians is the centroid of a triangular area.)

Illustration 4. Prove that the diagonals of a rhombus are perpendicular

Solution. Choose axes so that two vertices are at $A(0, 0)$ and $B(a, 0)$ (Fig. 24). Now a rhombus is an equilateral parallelogram and the coordinates of C and D must be such that $AC = CD = BD = AB = a$. In the figure we have purposely omitted writing the coordinates of C and D to emphasize that extreme care must be taken in doing so. Since $AC = a$, the coordinates of C may be written in terms of a and θ, the inclination of AC, thus: $C(a \cos \theta, a \sin \theta)$. Similarly for $D\ (a + a \cos \theta, a \sin \theta)$. Then

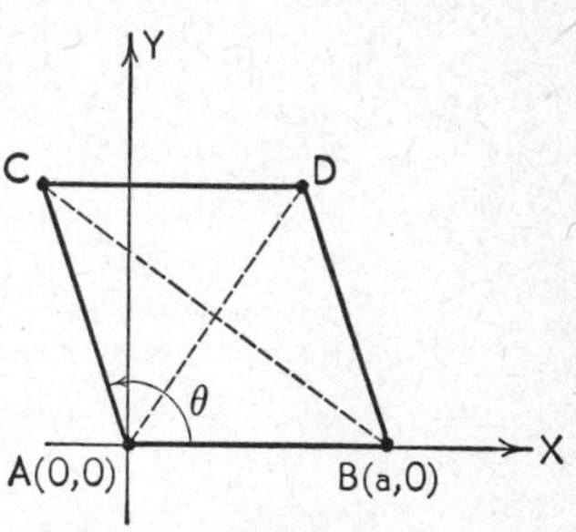

FIG. 24

$$m_{AD} = \frac{a \sin \theta}{a + a \cos \theta} = \frac{\sin \theta}{1 + \cos \theta}.$$

(Incidentally $\dfrac{\sin \theta}{1 + \cos \theta} = \tan \dfrac{\theta}{2}$ and the inclination of AD is $\theta/2$, which proves, in passing, that the diagonal AD of a rhombus bisects interior angle BAC.) Again

$$\begin{aligned} m_{BC} &= \frac{a \sin \theta}{a \cos \theta - a} = \frac{\sin \theta}{\cos \theta - 1} = \frac{\sin \theta}{\cos \theta - 1} \cdot \frac{\cos \theta + 1}{\cos \theta + 1} \\ &= \frac{\sin \theta(\cos \theta + 1)}{\cos^2 \theta - 1} = \frac{\sin \theta(\cos \theta + 1)}{-\sin^2 \theta} \\ &= -\frac{\cos \theta + 1}{\sin \theta}. \end{aligned}$$

Since $m_{AD} = -\dfrac{1}{m_{BC}}$, the diagonals are perpendicular.

Illustration 5. As an exercise develop the formulae for (a) distance and (b) slope with respect to a system of skewed (oblique) axes.

Solution. (a) Let the axes make an acute angle θ as in Fig. 25. The fundamental triangle P_1P_2Q is not a right triangle but has legs $x_2 - x_1$ and $y_2 - y_1$ parallel to the axes. The angle $P_1QP_2 = 180° - \theta$. We now make use of the law of cosines $\overline{P_1P_2}^2 = \overline{P_1Q}^2 + \overline{QP_2}^2 - 2(P_1Q)(QP_2) \cos (180° - \theta)$. (See (9), (b), p. 8.) Calling $d = \overline{P_1P_2}$ and noting that $\cos (180° - \theta) = -\cos \theta$, we get

$$(1) \quad d = \sqrt{(x_2 - x_1)^2 + 2(x_2 - x_1)(y_2 - y_1) \cos \theta + (y_2 - y_1)^2}.$$

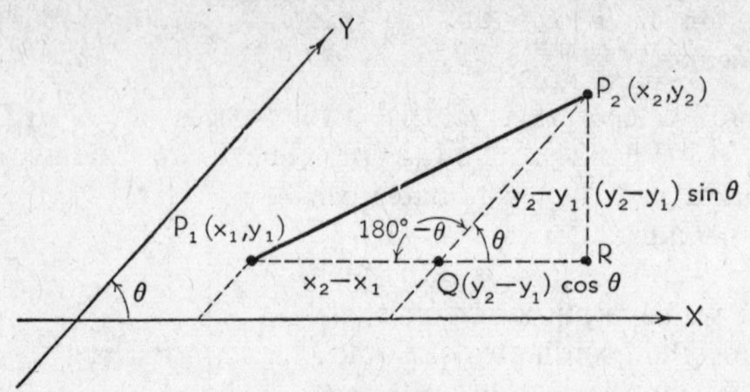

FIG. 25

In case $\theta = 90°$ the middle term drops out with $\cos 90° = 0$ and this formula reduces to that developed previously for rectangular coordinates.

(b) Project QP_2 in the x direction into QR and in a direction perpendicular to the X-axis into RP_2. It is clear that $QR = (y_2 - y_1) \cos \theta$ and $RP_2 = (y_2 - y_1) \sin \theta$. Then the slope of P_1P_2 will be given by

$$m = \frac{(y_2 - y_1) \sin \theta}{(x_2 - x_1) + (y_2 - y_1) \cos \theta}. \tag{2}$$

These formulae are more complicated than the corresponding ones in rectangular coordinates. The concept of slope naturally involves perpendicularity, and distance is most simply computed when the fundamental triangle is right.

Finally we point out that oblique axes and formula (2) could readily be used to prove the proposition of Illustration 4. Oblique axes could be chosen so that the vertices of the rhombus are $A(0, 0)$, $B(a, 0)$, $C(0, a)$, and $D(a, a)$, which is simpler than with rectangular axes. Then the slopes of AD and BC are as already determined and the work proceeds as before.

16. Accuracy in Drawings. Throughout his study of analytic geometry the student should make every attempt to make careful and accurate drawings to accompany his algebraic work. A figure that is reasonably exact will often lend aid to the analytic imagination whereas a poorly drawn figure may lead the student to make inaccurate deductions.

EXERCISES

1. For the triangle $A(1, 3)$, $B(-2, 1)$, $C(0, -4)$ find

(a) Distance BC. *Ans.* $\sqrt{29}$.

(b) Slope AB. *Ans.* $\frac{2}{3}$.

(c) Slope of a line perpendicular to AB. *Ans.* $-\frac{3}{2}$.
(d) Midpoint of AC. *Ans.* $(\frac{1}{2}, -\frac{1}{2})$.
(e) Coordinates of the centroid *Ans.* $(-\frac{1}{3}, 0)$.
(f) Direction cosines of the directed side AB. *Ans.* $\dfrac{-3}{\sqrt{13}}, \dfrac{-2}{\sqrt{13}}$.
(g) Slope of a line parallel to AC. *Ans.* 7.
(h) Angle ABC. *Ans.* $\tan\theta = \frac{19}{4}$.
(i) Area of ABC. *Ans.* $\frac{19}{2}$ sq. units.

2. Prove the following theorems by analytic methods:
(a) The midpoint of the hypotenuse of a right triangle is equidistant from the vertices.
(b) The line segment joining the midpoints of two sides of a triangle is parallel to the third side and equal to one-half its length.
(c) The lines joining the midpoints of the sides of a triangle divide it into four equal triangles.
(d) The distance between the midpoints of the non-parallel sides of a trapezoid is one-half the sum of the parallel sides.
(e) The diagonals of a trapezoid are equal if and only if the trapezoid is isosceles.
(f) The line segments joining the midpoints of adjacent sides of a quadrilateral form a parallelogram.
(g) The sum of the squares on the sides of a parallelogram equals the sum of the squares on the diagonals.

CHAPTER III

EQUATIONS AND GRAPHS

17. Basic Definitions. Numbers plotted as distances from a fixed point on a line such as the X-axis are called *real numbers.* The *absolute value* of a real number a, written $|a|$, is the positive magnitude of a. Thus $|7| = |-7| = 7$.

A collection of objects is called a *set.* A *relation* is a set of ordered pairs. The order in a given pair (x, y) is, first, the *element* x, second, the *element* y. Examples of relations are: (i) the three ordered pairs $(2, 3)$, $(2, 4)$, $(3, -7)$; (ii) the set of all ordered pairs of the form (x, y) where x is any real number and where y satisfies $y^2 = x^4$.

A function is a set of ordered pairs such that no two ordered pairs have the same first element. Examples of functions are: (iii) the one ordered pair $(2, 3)$; (iv) the set of all ordered pairs of the form (x, x^2-2), where x is any real number.

The set X, of all first elements x, is called the *domain* of the relation (or function) and the set Y, of all second elements y, is called the *range* of the relation (or function). In the examples above, the domain and range are:

Example	Domain	Range
(i)	The two numbers 2, 3	The three numbers 3, 4, -7
(ii)	Every real number	Every real number
(iii)	2	3
(iv)	Every real number	Every number ≥ -2

Common notations for a function are: (a) The set (x, y) where $y = f(x)$; (b) f: (x, y). It is common to call x the *independent variable* and y the *dependent variable*; also y, or $f(x)$, is called the *value* of f at x.

A given rule determines y for each value of x. If a function is determined by $y = f(x)$, then this equation is the rule.

A special function of interest is the "absolute value" function which is the set of ordered pairs (x, y) where $y = |x|$, defined as follows:

$$y = |x| = \begin{cases} x, \text{ if } 0 \leq x \\ -x, \text{ if } x < 0 \end{cases}$$

The inverse function f^{-1}, inverse to f, is defined to be the set of ordered pairs (y, x) (no two of which have the same first element y). Thus the function (2, 3) has the inverse (3, 2); but the function (3, 7), (4, 7) has no inverse. The function f, where $y = 2x - 1$, has the inverse f^{-1}, where $x = \frac{y + 1.}{2}$ This follows since if we solve $y = 2x-1$ for x, we get $x = \frac{y + 1.}{2}$ Often we want to write a function and its inverse in terms of the same letter x. In the example we would therefore say that the function f: $(x, 2x-1)$ has the inverse f^{-1}: $\left(x, \frac{x + 1}{2}\right)$. Therefore, if we are given $y = f(x)$, to find the inverse function, we solve for $x = f^{-1}(y)$ first of all, and then, if we wish to have the usual variables x (independent) and y (dependent), we switch letters, writing $y = f^{-1}(x)$.

The graph of a function (or relation) is obtained by plotting the set of ordered pairs.

The function f is *periodic* of period P if $f(x + P) = f(x)$.

The value $f(x)$ is said to approach b as a limit, when x approaches a, if the value of $|f(x) + b|$ becomes and remains less than any preassigned quantity. This is usually written

$$\lim_{x \to a} f(x) = b$$

and is read "the limiting value of f, as x approaches a, equals b." A function f is said to be *continuous at the point a* if three conditions are satisfied, namely: (i) $\lim_{x \to a} f(x)$ exists, (ii) $f(a)$ exists, and (iii) $\lim_{x \to a} f(x) = f(a)$. For example, the function f where $y = x^2$ is a continuous function at the point $x = 0$ since $\lim_{x \to 0} x^2 = 0 = f(0)$. If $\lim_{x \to a} f(x) \neq f(a)$, the function is said to be *discontinuous at the point a.* For example, the function f, defined as follows, is discontinuous at $x = 0$: $f(x) = x^2$, for values of $x \neq 0$, but $f(0) = 1$. For $\lim_{x \to 0} x^2 = 0$ whereas $f(0) \neq 0$. Or again the function $f = 1/x$ is discontinuous at the origin since $\lim_{x \to 0} 1/x$ does not exist (it is infinite).

A function is *continuous in an interval* if it is continuous at every point in the interval.

18. Equations. An *equation* $F(x) = 0$ imposes a condition on the variable x which then can assume only certain values. For example, if the equation is $ax^2 + bx + c = 0$, x can take on only two values, namely $x = \dfrac{-b \pm \sqrt{b^2 - 4ac}}{2a}$. The values of x which satisfy $F(x) = 0$ are said to be the solutions or *roots* of the equation. All of the real solutions of $F(x) = 0$ may be represented by points on a line, the X-axis. These points would then constitute the *graph* of the equation in one dimension.

For example, the graph, in one dimension, of $x - k = 0$ would be the one point located k units from the origin of the X-axis. It is also simple to interpret $x - k = 0$ as a graph in two dimensions, for x always equals k, and y can be arbitrary since its value does not affect the equation. Hence every point (k, a), regardless of the value of a, is a point on the graph which is, consequently, a straight line perpendicular to the X-axis through $(k, 0)$. Thus the graph of $F(x) = 0$ in two dimensions will be a set of straight lines erected perpendicular to the X-axis, one at each root of $F(x) = 0$.

An equation $F(x, y) = 0$ might be solved for y in terms of x, say $y = f(x)$, indicating that x was to be considered the independent and y the dependent variable.

This is one of the central problems in plane analytic geometry: *given a function f determined by* $y = f(x)$, *to plot its graph* or to represent it geometrically. We sometimes say that the graph of f is the *locus* of f. The curve traced by a moving point is called the locus of the point.

Of course there is much more to plotting than just putting down points here and there. By a thorough study of the *equation* much can be learned about the geometric properties of its *graph.* Such an analysis is one of the roles of analytic geometry.

19. Discussion of Equations and Their Graphs. In the study of an equation $y = f(x)$ there are four principal analyses to be made in order to be able to plot the graph with confidence and some accuracy. These are a determination of the: I, *intercepts;* II, *extent;* III, *symmetry;* IV, *asymptotes.*

I. *Intercepts.* The *intercepts* are the points where the curve crosses the axes. The x-intercepts are obtained by setting $y = 0$ and solving for x. Such a value of x — for which y, the function, is zero — is called a *zero* of the function f; it is, of course, a *root* of the equation $f(x) = 0$. The y-intercepts are gotten by setting $x = 0$ and solving for y.

Illustration 1. Find the intercepts of the graph of $y = x^2 - 3x + 2$.

Solution. Setting $y = 0$ the roots of $x^2 - 3x + 2 = 0$ are determined as $x = 1$ and $x = 2$. Hence the x-intercepts are $(1, 0)$ and $(2, 0)$. Similarly setting $x = 0$ we get $y = 2$; hence the one y-intercept has coordinates $(0, 2)$.

II. *Extent.* In plotting the graph of an equation it is useful to know the *extent* of the graph. The graph might or might not be confined to finite regions of the plane. If the equation is of the form $y = f(x)$, the independent variable x can range from $-\infty$ to $+\infty$. But the values of x in certain regions (intervals) may lead to complex values of y, and there would then be no corresponding graph in those regions. Or again the y values might be real but bounded in one direction so that no y would exceed (or be less than) a certain number M, in which case the graph would extend only throughout half of the plane. Obvious modifications in the discussion would permit these ideas to be applicable to boundedness in the direction of any coordinate axis, positive or negative.

Illustration 2. Discuss the extent of the graph of $y = x^2 - 3x + 2$.

Solution. The independent variable x may range from $-\infty$ to $+\infty$. To find out whether y is bounded we solve the equation for x as a function of y. We have

$$x^2 - 3x + (2 - y) = 0,$$

which yields

$$x = \frac{3 \pm \sqrt{9 - 4(2 - y)}}{2},$$
$$= \tfrac{3}{2} \pm \tfrac{1}{2}\sqrt{1 + 4y}.$$

From this it is evident that y is restricted to values not less than $-\frac{1}{4}$ since, if $y < -\frac{1}{4}$, x would then be complex. The whole graph lies in the upper half plane determined by the line through $(\frac{3}{2}, -\frac{1}{4})$ parallel to the X-axis. The point $(\frac{3}{2}, -\frac{1}{4})$ is the minimum point on the graph.

Illustration 3. Discuss the extent of the graph of the equation $y^2 = x(x - 1)(x + 3)$.

Solution. First note that for a given x there are two values of y, namely $+y$ and $-y$. The right-hand side of the equation must be positive since the left-hand side is always positive. This immediately tells us that there is no part of the graph in the regions for which $x < -3$ and $0 < x < 1$. As $x \to +\infty$ (read "as x approaches infinity"), $y \to \pm\infty$ as can be seen directly from the equation. Unfortunately the quadratic equation methods of Illustration 2 are not readily extensible to this present equation, so that we cannot easily determine the extent of y in the interval $-3 < x < 0$. But we may reason as follows: for $x = -3$ and $x = 0$, y is zero; and for no x in the interval $-3 < x < 0$ will y become infinite. Therefore y is bounded above and below. (By advanced methods of the calculus it can be shown that the maximum and minimum values of y are $\pm\sqrt{\dfrac{70 + 13\sqrt{52}}{27}}$ respectively and that they occur at $x = -\frac{2}{3} - \frac{1}{6}\sqrt{52}$.)

III. *Symmetry.* The points (x, y) and $(x, -y)$ are *symmetric with respect to the X-axis,* the one being the "mirror image" of the other. Either point is said to be the *reflection* of the other *about the X-axis.* For the purposes of general discussion it is best to consider the equation of a graph in the form $F(x, y) = 0$. It should be evident that the graph will be symmetric with respect to the X-axis if $F(x, y) \equiv F(x, -y)$, since then if (x, y) is a point on the graph $[F(x, y) = 0]$, so also will $(x, -y)$ be a point on the graph $[F(x, -y) = 0]$. Similarly, if $F(x, y) \equiv F(-x, y)$, the curve will be symmetric with respect to the Y-axis. Further, since a line joining (x, y) and

$(-x, -y)$ passes through the origin and the distance from (x, y) to the origin is the same as the distance from $(-x, -y)$ to the origin, the graph will be *symmetric with respect to the origin* if $F(x, y) \equiv F(-x, -y)$. Either point is said to be the *reflection* of the other *about the origin.* If there is symmetry with respect to both axes there is, necessarily, symmetry with respect to the origin, but not conversely.

These are special cases of *symmetry with respect to a line* and *symmetry with respect to a point.* In only a few cases shall we extend the discussion beyond that for the axes and the origin.

Illustration 4. Examine $F(x, y) \equiv x^2 - x + y^4 - 2y^2 - 6 = 0$ for symmetry.

Solution. Since y enters with even powers only, it is clear that changing y into $-y$ will not affect the function F. Hence there is symmetry with respect to the X-axis. If, however, $-x$ is substituted for x, F is changed to $x^2 + x + y^4 - 2y^2 - 6$. Since $F(x, y) \not\equiv F(-x, y)$ there is no symmetry with respect to the Y-axis. There is also no symmetry with respect to the origin.

Illustration 5. Examine the following for symmetry.

(a) $y^2 = x(x - 1)(x + 3)$, (Illustration 3)
(b) $F(x, y) \equiv x^2 - y^2 - 3 = 0$,
(c) $G(x, y) \equiv xy - 1 = 0$.

Solution.

(a) Symmetric with respect to the X-axis since y enters only to an even power. There is no other symmetry.
(b) Here $F(x, y) \equiv F(x, -y)$, $F(x, y) \equiv F(-x, y)$, and $F(x, y) \equiv F(-x, -y)$. Hence there is symmetry with respect to both axes and to the origin.
(c) $G(x, y) \equiv G(-x, -y)$ and there is symmetry with respect to the origin.

Summary of Tests for Symmetry

The graph of the equation $F(x, y) = 0$ will be symmetric with respect to

1. The X-axis if $F(x, y) \equiv F(x, -y)$. (Fig. 26)
2. The Y-axis if $F(x, y) \equiv F(-x, y)$. (Fig. 27)
3. The origin if $F(x, y) \equiv F(-x, -y)$. (Fig. 28)

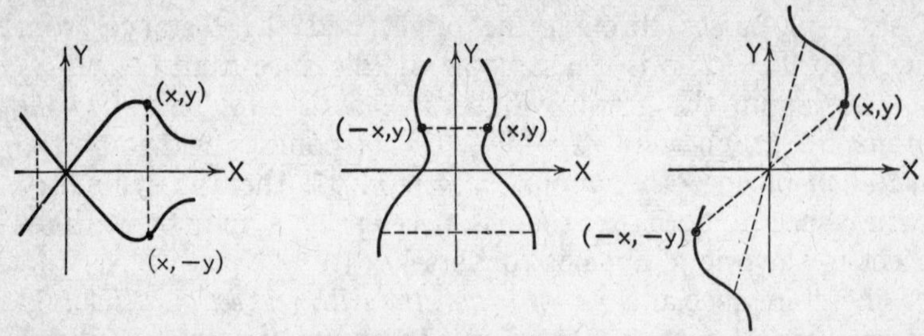

FIG. 26. Symmetry with respect to X-axis. FIG. 27. Symmetry with respect to Y-axis. FIG. 28. Symmetry with respect to the origin.

IV. *Asymptotes.* It is of interest to study the behavior of an unbounded curve in the neighborhood of infinity, where either x, or y, or both become infinite. A curve may recede to infinity in a certain direction, and if this direction can be determined it will be of great aid in plotting the curve.

Let the curve be given by $y = f(x)$ and let $x = g(y)$ be the form of the equation when solved for x as a function of y

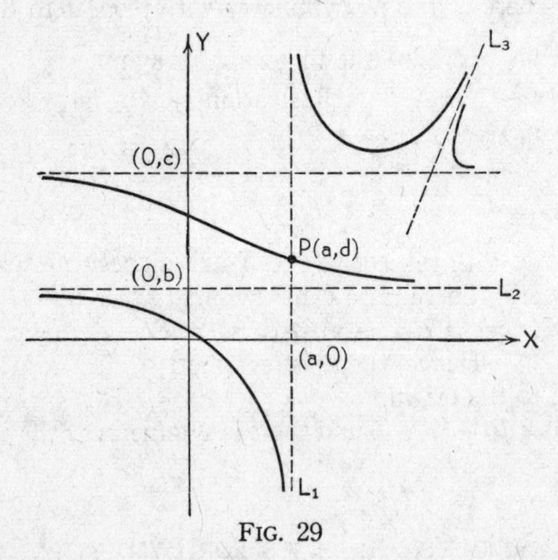

FIG. 29

(Fig. 29). Since we are essentially interested here in the directions parallel to the axes, we lay down the following special definitions:

(1) The line L_1 through $(a, 0)$ parallel to the Y-axis is called a *vertical asymptote* if, as $x \to a$, $|y| \to \infty$.

This can be written

(1′) $x = a$ is a vertical asymptote if $\lim_{x \to a} f(x) = \pm\infty$.

(2) The line L_2 through $(0, b)$ parallel to the X-axis is called a *horizontal asymptote* if, as $y \to b$, $|x| \to \infty$.

This can be written

(2′) $y = b$ is a horizontal asymptote if $\lim_{y \to b} g(y) = \pm\infty$.

A curve with such asymptotic lines is necessarily discontinuous at $x = a$ and at $y = b$ and the curve is made up of several separate pieces. It is important to remark that $f(x)$ might be multiple-valued and therefore have a finite as well as an infinite determination for a given value of x. This is illustrated in the figure by the point P; one determination of y is given by $f(a) = d$ — the point P — and another is $f(a) = \pm\infty$. A similar remark applies to $g(y)$. Also a curve might have several asymptotes of either variety and others as well in directions not parallel to an axis. Without going into detail we say simply that if a curve approaches indefinitely near any straight line in the neighborhood of infinity, then that straight line is an asymptote. This is illustrated by the line L_3 in the figure.

We now combine these analyses in several illustrations.

Illustration 6. Find the intercepts, discuss extent and symmetry, find the vertical and horizontal asymptotes, and sketch the graph of

$$y = \frac{x(x-1)}{x+2}.$$

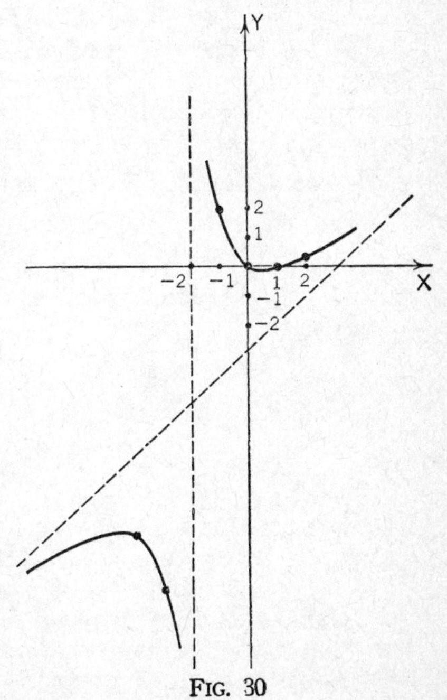

FIG. 30

Solution. I. *Intercepts.* The numerator of $f(x)$ is zero when $x = 0, 1$. The x-intercepts are, therefore, $(0, 0)$ and $(1, 0)$. Since, when $x = 0$, y has only the value zero, there is no y-intercept except $(0, 0)$.

II. *Extent.* The graph is not confined to a finite portion of the plane since both x and y can range from $-\infty$ to $+\infty$.

III. *Symmetry.* The tests of symmetry indicate there is no symmetry with respect to either the axes or the origin.

IV. *Asymptotes.* From the denominator of $f(x)$ we note that, as $x \to -2$, $y \to \pm\infty$. Therefore the line parallel to the Y-axis passing through $(-2, 0)$, $x = -2$, is a vertical asymptote. There is no finite value of y for which x is infinite; therefore there is no horizontal asymptote. But we note that the ratio $\frac{x(x-1)}{x+2}$ approaches the value x as x gets bigger and bigger since $\frac{x-1}{x+2}$ approaches unity. Therefore there is an asymptotic line with inclination 45° but we are unable at this time to determine the exact position of this line.

Looking at the function we see that it is negative for $x < -2$ and for $0 < x < 1$; it is positive for $-2 < x < 0$ and for $x > 1$. The essential shape of the curve is consequently determined without plotting any specific points other than $(0, 0)$ and $(1, 0)$. We compute the coordinates of a few points such as $(2, \frac{1}{2})$, $(\frac{1}{2}, -\frac{1}{10})$, $(-1, 2)$, $(-3, -12)$, $(-4, -10)$ and proceed to sketch the graph.

Illustration 7. Analyze and sketch the graph of

$$y \equiv f(x) = \frac{(x+1)(x-3)}{x^2 - 4}.$$

Solution. In the process of solving this relation for x as a function of y we get

$$x^2(y-1) + 2x + (3 - 4y) = 0, \tag{1}$$

from which it follows that

$$x \equiv g(y) = \frac{-1 \pm \sqrt{4y^2 - 7y + 4}}{y - 1}. \tag{2}$$

I. *Intercepts.* The x-intercepts are computed to be $(-1, 0)$ and $(3, 0)$. When $x = 0$, $y = \frac{3}{4}$. Therefore the y-intercept is $(0, \frac{3}{4})$.

II. *Extent.* Since both x and y can range freely from $-\infty$ to $+\infty$ the graph is not bounded by any finite portion of the plane.

III. *Symmetry.* The curve is symmetric with respect neither to the axes nor to the origin since the tests for such symmetry fail.

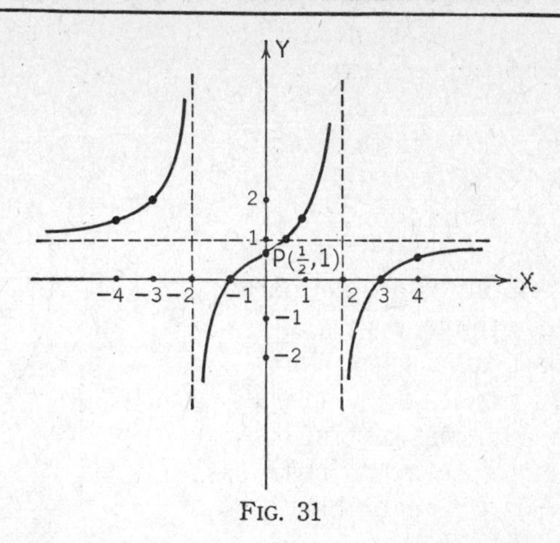

FIG. 31

IV. *Asymptotes.* The denominator of $f(x)$ is $x^2 - 4$; setting this equal to zero yields vertical asymptotes through $(2, 0)$ and $(-2, 0)$. The denominator of $g(y)$ is $y - 1$. For $y = 1$, x, from (2), is $x = \dfrac{-1 \pm 1}{0}$. Using the plus sign, x takes the form $x = \frac{0}{0}$, which is indeterminate. But from (1) we see that when $y = 1$ there is a finite value of x, $x = \frac{1}{2}$. [Equation (1), normally a quadratic in x, reduces to a linear equation when $y = 1$ since the coefficient of x^2 is then zero. A quadratic equation $ax^2 + bx + c = 0$ may be thought of as having one infinite root and one finite root, namely $x = -c/b$, if $a = 0$.] Thus there is one horizontal asymptote passing through $(0, 1)$. We have determined that not only is $|x| = \infty$ when $y = 1$ but also that $x = \frac{1}{2}$ when $y = 1$. Thus one branch of the curve crosses an asymptote, and the point $P(\frac{1}{2}, 1)$ corresponds to the point P as previously discussed and as indicated in Fig. 29.

We now compute the coordinates of a few points: $(4, \frac{5}{12})$, $(1, \frac{4}{3})$, $(-3, \frac{12}{5})$, $(-4, \frac{21}{12})$. It is simple to sketch the graph as in Fig. 31.

Illustration 8. Analyze and sketch the graph of $y^2 = x(x-1)(x+3)$ (Illustration 3). See Fig. 32, p. 38.

Solution. The intercepts are $(-3, 0)$, $(0, 0)$, and $(1, 0)$. The curve is bounded as indicated in Illustration 3. There is symmetry with respect to the X-axis. There are no asymptotes. A few points on the graph are $(-2, \pm\sqrt{6})$, $(-1, \pm 2)$, $(2, \pm\sqrt{10})$, $(3, \pm 6)$.

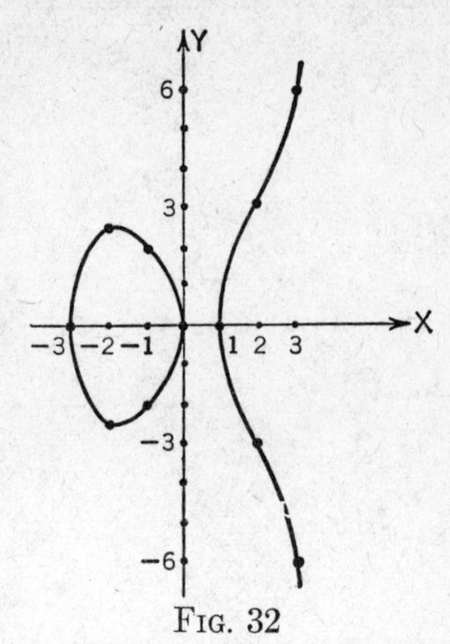

Fig. 32

It will be noticed that in the graphs of these illustrations the asymptotic lines are dotted in. Accurate graphs cannot readily be constructed without the presence of the asymptotes, where they exist; but it should be clear that the asymptotes are *not* part of the above graphs. Asymptotes *could* be part of a graph, however, as will result as a special case of the following general discussion of factorable equations.

If $F(x, y)$ is factorable let $F(x, y) \equiv G(x, y)H(x, y)$. It is obvious that any number pair (x, y) which makes $G = 0$ will also make $F = 0$. Therefore $G = 0$ is a part of the graph of $F = 0$. Likewise any point on $H = 0$ is also on $F = 0$. Hence $F = 0$ is made up of the separate graphs of $G = 0$ and $H = 0$ and the idea is extensible to any number of factors. If $G = 0$ is the equation of the asymptotes of $H = 0$, then $F = 0$ is a graph containing its asymptotes.

Illustration 9. Sketch

$$(y - 1)(x^2 - 4)[y(x^2 - 4) - (x + 1)(x - 3)] = 0.$$

Solution. The last factor is the function $F(x, y)$ involved in Illustration 7. Its graph is shown in Fig. 31. Since the first two factors yield the asymptotes, the total graph is made up of Fig. 31 plus the asymptotes.

Illustration 10. Sketch $F(x, y) \equiv x^2 - y^2 = 0.$

Solution. Since F can be factored into $F(x, y) \equiv (x - y)(x + y)$ its graph will be made up of the graphs of $x - y = 0$ and $x + y = 0$. The first of these equations states that $x = y$; hence it is a straight line making an angle of 45° with the positive direction of the X-axis and obviously passing through the origin. Similarly the second equation says that $x = -y$, which is a straight line of inclination 135° and also passing through the origin. These two lines constitute the graph.

There are other fundamental questions that arise in connection with plotting or *curve tracing* as it is often called, but most of them involve at least a knowledge of the calculus.

20. Intersection of Curves. The rectangular coordinates of a point common to the graphs of two equations will satisfy simultaneously both equations, and the coordinates of no other point will do so. Such a point is called a point of intersection of the curves. Hence to find the points of intersection of two curves we solve simultaneously the equations of the curves and pair properly the resulting values of x and y.

Illustration 1. Find the points of intersection of the curves (a) $y = x^2$ and (b) $x - y + 2 = 0$.

Solution. In solving (a) and (b) simultaneously we simply substitute the value of y from either equation into the other and then proceed to solve the resultant quadratic equation in x. Thus

$$x - x^2 + 2 = 0,$$

whence

$$x = -1, 2,$$

and

$$y = 1, 4.$$

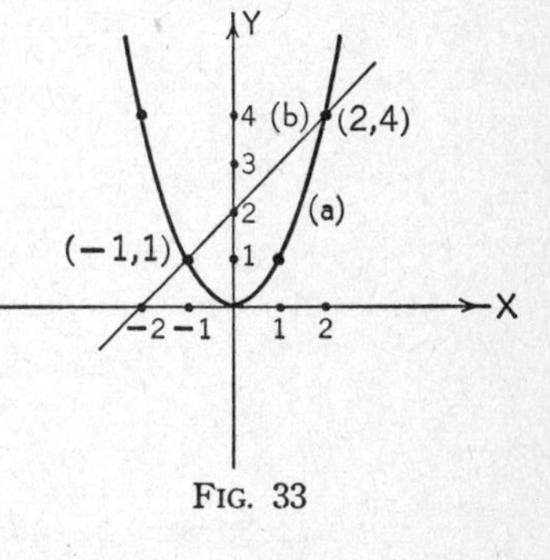

FIG. 33

These must be paired properly to give the coordinates of the points of intersection $(-1, 1)$ and $(2, 4)$. [Equation (a) plots a parabola, (b) a straight line.]

Illustration 2. Sketch and find the points of intersection of (a) $x^2 + 4y^2 = 4$ and (b) $x^2 - y^2 = 1$.

Solution. (a) Intercepts are $(\pm 2, 0)$, $(0, \pm 1)$. The curve is bounded by $-2 \leqq x \leqq 2$, $-1 \leqq y \leqq 1$. There is symmetry with respect to both axes and to the origin. The curve is closed and there are no asymptotes. A few points are plotted showing an oval-shaped graph. (It is an ellipse.)

(b) Intercepts are $(\pm 1, 0)$. For $-1 < x < 1$, y is complex so there is no part of the graph in this region. Otherwise x can range to $\pm\infty$ and the graph has two branches. There is symmetry with respect to both axes and to the origin. There are two asymptotes, as indicated in Fig. 34, but we cannot at this time show this analytically. (The curve is a hyperbola.)

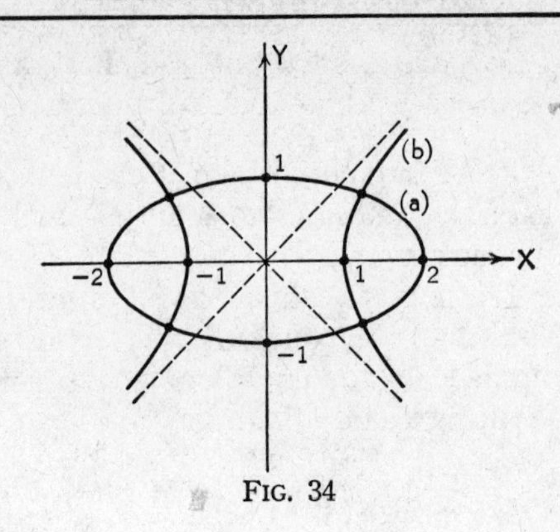

FIG. 34

From (b) $x^2 = 1 + y^2$. Substituting this in (a) yields

$$\begin{aligned} 1 + y^2 + 4\,y^2 &= 4, \\ 5\,y^2 &= 3, \\ y &= \pm\sqrt{\tfrac{3}{5}}. \end{aligned}$$

The corresponding values of x are given by substituting these values of y back into either equation. From (b)

$$\begin{aligned} x^2 &= 1 + y^2, \\ &= \tfrac{8}{5}, \\ x &= \pm\sqrt{\tfrac{8}{5}} = \pm 2\sqrt{\tfrac{2}{5}}. \end{aligned}$$

Hence the points of intersection are four in number, namely $(2\sqrt{\tfrac{2}{5}}, \sqrt{\tfrac{3}{5}})$, $(2\sqrt{\tfrac{2}{5}}, -\sqrt{\tfrac{3}{5}})$, $(-2\sqrt{\tfrac{2}{5}}, \sqrt{\tfrac{3}{5}})$, $(-2\sqrt{\tfrac{2}{5}}, -\sqrt{\tfrac{3}{5}})$.

EXERCISES

Trace the following curves.

1. $x^2 + y^2 = 1$.
2. $2\,x - 3\,y + 1 = 0$.
3. $4\,x^2 - y^2 = 1$.
4. $y^2 = x$.
5. $y = (x - 1)(x + 2)(x + 3)$.
6. $y = \dfrac{x(x - 2)}{x + 3}$.
7. $y^2 = x(x^2 - 4)$.
8. $x^2 + y^2 = x^3$.
9. $y^2 = x^3$.
10. $y = x^4$.

Find the points of intersection of the following curves and sketch.

11. (a) $x^2 + y^2 = 1$, (b) $2x - 2y = 1$.

12. (a) $y = x^2$, (b) $2x - y - 1 = 0$.

13. (a) $y^2 = x^3$, (b) $x = 2$.

14. (a) $y^2 = x$, (b) $y^2 = 1 - x$.

15. (a) $y^2 = x$, (b) $x^2 = y$.

21. Loci. We have already stated (p. 30) that one of the central problems in plane analytic geometry is to discuss and plot a given function $y = f(x)$. This implies an algebraic study of the given function (or equation). Such studies have been made in a number of illustrations.

We now come to the second central problem in analytic geometry: *given a curve, defined by certain geometric conditions, to find its equation.* This is generally known as a *locus* problem: *to find the locus of a point which moves according to some prescribed law.* To find a locus means to find the equation of the locus and to analyze, sketch, and identify it if possible.

Until we have made some systematic study of certain standard curves we cannot take up many problems on loci. We give at this time only a few simple illustrations, reserving a fuller discussion for Chapters XIV, XV, and XVI.

Illustration 1. Find the locus of a point which moves so that it is always equidistant from the two fixed points $A(1, 2)$ and $B(-1, 0)$.

Solution. We pick out, by sight, some point P in the plane which (approximately) satisfies the conditions of the problem and give to P the general coordinates (x, y). Any true relation that can be found connecting the variables x, y will contain the equation of the locus. It will contain nothing more provided no extraneous condition is introduced.

From the formula for the distance between two points we may write

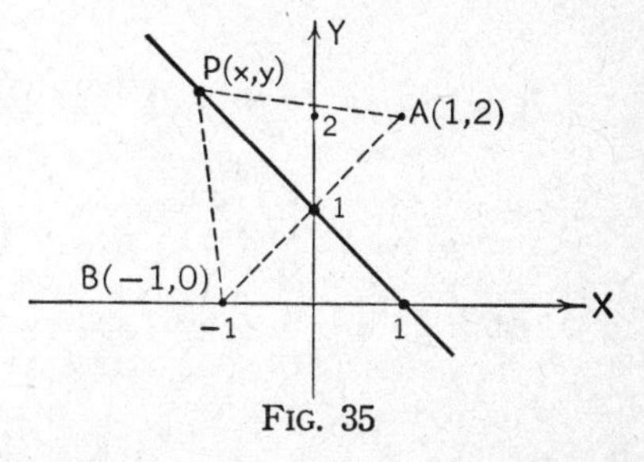

FIG. 35

(a) $\sqrt{(x-1)^2 + (y-2)^2}$
$= \sqrt{(x+1)^2 + y^2}$

and this is the equation of the locus. It is better, however, to rationalize this equation by squaring both sides. Thus

$$x^2 - 2x + 1 + y^2 - 4y + 4 = x^2 + 2x + 1 + y^2,$$

which reduces to

(b) $$x + y - 1 = 0.$$

We know from plane geometry that the locus is the line perpendicular to AB and bisecting it.

Illustration 2. A point moves so as always to be a constant distance r from the origin. Find its locus.

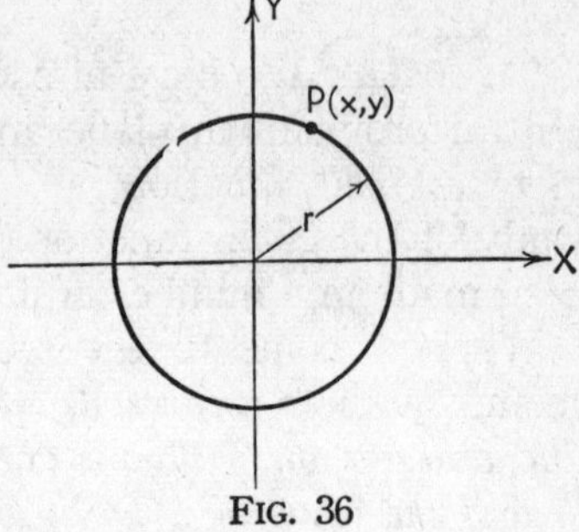

FIG. 36

Solution. The distance from $P(x, y)$ to $(0, 0)$ is $\sqrt{x^2 + y^2}$ and this must be equal to r. Hence the locus is given by

$$x^2 + y^2 = r^2$$

and is a circle of radius r with center at the origin.

Illustration 3. Find the locus of a point which moves so that it is always twice as far from the X-axis as it is from the Y-axis.

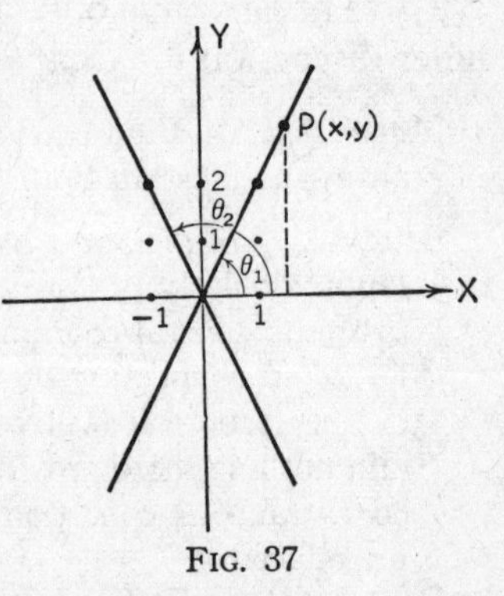

FIG. 37

Solution. Let a general point on the locus be $P(x, y)$. The conditions of the problem, when *distance* is interpreted as positive, imply both $y = 2x$ and $y = -2x$. Hence the equation of the locus must contain both of these equations and is

(1) $(y - 2x)(y + 2x) = 0,$

or $y^2 - 4x^2 = 0.$

From (1) it is evident that the locus is two straight lines through the origin with inclinations θ_1 and θ_2 given by $\tan\theta_1 = 2$ and $\tan\theta_2 = -2$.

Illustration 4. A line segment AB of length L moves so that A always lies on the X-axis and B on the Y-axis. Find the locus of the midpoint of AB.

Solution. Call the coordinates of the midpoint $P(x, y)$. Then A and B have coordinates $A(2x, 0)$ and $B(0, 2y)$. Making use of the Pythagorean theorem we have

$$(2\,x)^2 + (2\,y)^2 = L^2,$$

or

$$x^2 + y^2 = \left(\frac{L}{2}\right)^2.$$

This shows that the locus is a circle with center at (0, 0) and radius $L/2$. (Compare with Illustration 2.)

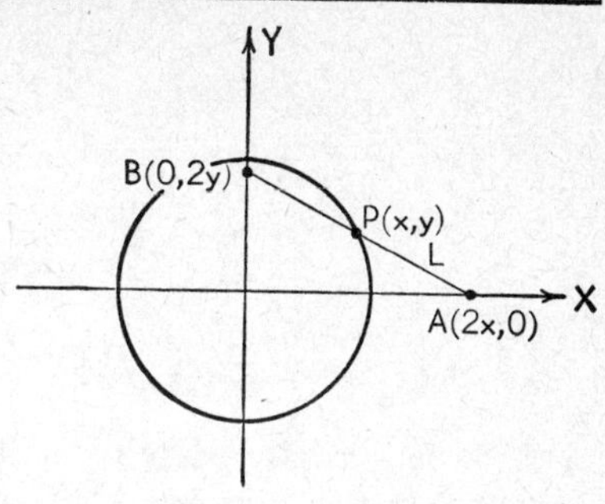

FIG. 38

Illustration 5. A variable circle passes through $A(2, 0)$ and is always tangent to the vertical line $x = -2$. Find the locus of the center of the circle.

Solution. Let $P(x, y)$ be the center and drop the perpendicular PB to the line $x = -2$. Now $PA = PB$; hence

$$\sqrt{(x-2)^2 + y^2} = x + 2,$$

which, upon squaring and reducing, gives as the desired locus

$$y^2 = 8\,x.$$

FIG. 39

The curve passes through (0, 0), lies in the right half plane, is symmetric with respect to the X-axis, and has no asymptotes. (It is a parabola, as we shall see in Chapter VI.)

EXERCISES

1. A point P moves so that the product of its ordinate and abscissa is a constant k. Find the locus of P. *Ans.* $xy = k$.

2. Given $A(0, 1)$ and $B(2, 5)$, two fixed points. Find the locus of P if the slope of AB equals that of BP. *Ans.* $2\,x - y + 1 = 0$.

3. P moves so that the absolute value of its distance from the horizontal line $y = 3$ is always 2 units. Find the locus of P. *Ans.* $(y-5)(y-1) = 0$.

4. The hypotenuse of a right triangle is the segment joining (0, 0) and (4, 0). Find the locus of the third vertex. *Ans.* $(x-2)^2 + y^2 = 4$.

5. Find the locus of a point which moves so that the sum of its distances from the two fixed points $F(4, 0)$, $F'(-4, 0)$ is 10 units.

Ans. $\dfrac{x^2}{25} + \dfrac{y^2}{9} = 1.$

CHAPTER IV

THE STRAIGHT LINE

22. Polynomials. Giving a few preliminary definitions we begin, with the straight line, a systematic study of certain standard curves. The simplest of these have polynomial equations.

A *polynomial in one variable* x is a function made up of a sum of terms each of the form Ax^N where A is a constant and N is a positive integer or zero. Thus $3x^3 - 5$, $2x^4 + 6x^2 - x$, and $\pi x^{17} + \sqrt{2}\,x^5 + \frac{7}{8}$ are polynomials. The general polynomial f is given by

$$(1) \qquad f(x) \equiv a_0x^n + a_1x^{n-1} + \cdots + a_{n-1}x + a_n.$$

The a's (coefficients) are constants. The polynomial is of the nth *degree* if $a_0 \neq 0$. In algebra it is proved that a polynomial of the nth degree has exactly n zeros. This amounts to saying that the equation $f(x) = 0$ has n roots. We have seen that $f(x) = 0$ plots n points in one dimension and n lines perpendicular to the X-axis in two dimensions if the n roots are real and distinct. A multiple root yields only one point (line), and complex roots do not plot.

A *polynomial in two variables* x and y is a function F where

$$(2) \quad F(x, y) \equiv a_0(y)x^n + a_1(y)x^{n-1} + \cdots + a_{n-1}(y)x + a_n(y),$$

where now the coefficients are themselves polynomials in y. It is of degree n in x if $a_0(y) \not\equiv 0$ and it is of degree m in y if some coefficient $a_i(y)$ is of degree m in y and no other coefficient is of greater degree. The *degree of a term* of F is the *sum of the degrees* in x and y, and the *degree of the polynomial* is the degree of the term of greatest degree. Thus $(y^2 - y)x^3 + y^6x^2 - x + y + 7$ is a polynomial of degree 3 in x, 6 in y; the degree of the polynomial is 8. The polynomial F determines an *algebraic function* of the two variables x and y.

23. The Linear Equation. The simplest polynomial in two variables (other than a constant, which we do not consider here) is one of the *first degree.* Such a polynomial is called a *linear function* and produces a *linear equation* when equated to zero:

$$Ax + By + C = 0. \tag{1}$$

We now prove

Theorem 1. The graph of a linear equation is a straight line.

Proof. First. We have already seen that $x = k$, $y = k$ represent straight lines parallel to the Y-axis and X-axis respectively.

Second. The graph of $y = mx$ is a straight line through (0, 0) with slope m since the ordinate of any point on this line is m times the abscissa and this is true for no point not on the line.

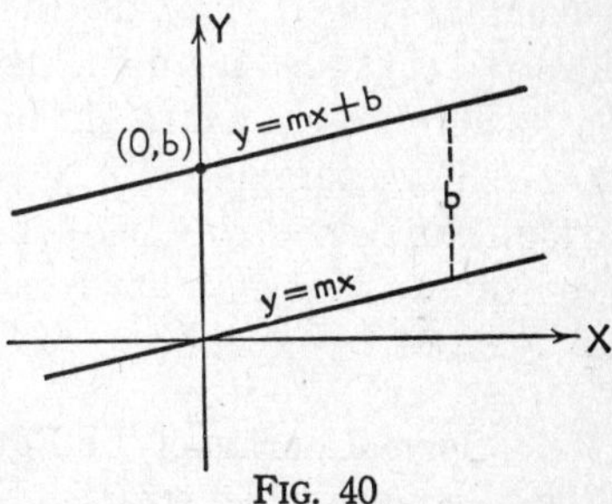

FIG. 40

Third. The graph of $y = mx + b$ contains the point $(0, b)$ and each ordinate exceeds the corresponding ordinate of $y = mx$ by the constant amount b. The graph therefore represents a line parallel to $y = mx$.

Fourth. Multiplying or dividing the equation $y = mx + b$ by a constant (not zero) will not affect the locus. Hence $By = Bmx + Bb$ plots the same locus and is of the form (1) where $A = -Bm$ and $C = -Bb$.

This completes the proof of the theorem.

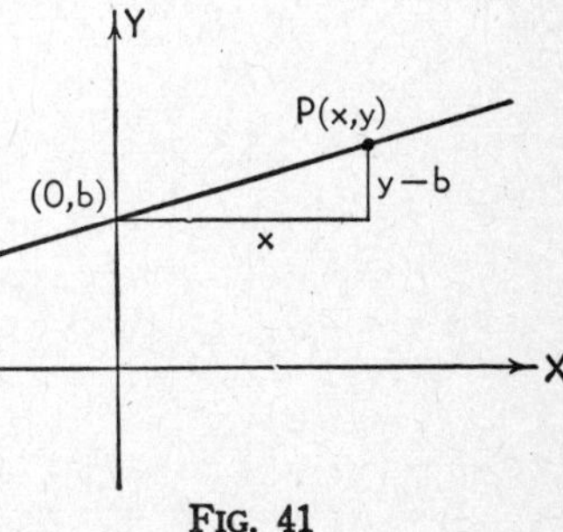

FIG. 41

Theorem 2. A straight line is represented by a linear equation.

Proof. First. A line parallel to the Y-axis (infinite slope) has a linear equation, namely $x = k$.

Second. Any other line will be determined by its slope m and one point on it, say the y-intercept $(0, b)$.

If $P(x, y)$ is any other point on the line, then $(y-b)/x=m$ and this relation holds for no point off the line. This relation reduces to $y = mx + b$, which is linear, and the proof is complete.

Note. Even though there is always a linear equation that represents a given line, equations of higher degree may also represent a straight line. For example $(Ax + By + C)^2 = 0$ represents a straight line since each factor of the left-hand side plots the *same* straight line. (There are, however, reasons why we should like to say that this equation represents *two* straight lines and that the lines coincide.) Or again $g(x, y)(Ax+By+C) = 0$ will represent a straight line if $g(x, y) = 0$ has no real point on it. For example $(x^2 + y^2 + 1)(Ax + By + C) = 0$ plots a straight line since $x^2 + y^2 + 1 = 0$ for no real values of x and y simultaneously. Similar remarks could be made about the equation of any graph.

24. Special Forms of the Equation of a Straight Line. Every (linear) equation of a straight line is of the form $Ax+By+C=0$; this is therefore called the *general equation* of a straight line. Certain special, or standard, forms of this equation are of both interest and use.

I. *Two-Point Form.* A line is determined by two distinct points on it, $P_1(x_1, y_1)$ and $P_2(x_2, y_2)$. Let $P(x, y)$ be any other point on the line. Then (and only then) by similar triangles

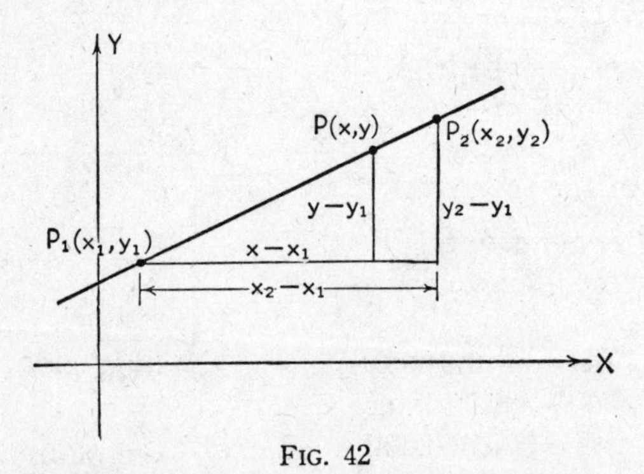

FIG. 42

(1) $$\frac{y - y_1}{x - x_1} = \frac{y_2 - y_1}{x_2 - x_1},$$

which is the desired equation.

II. *Point-Slope Form.* Since $m = \dfrac{y_2 - y_1}{x_2 - x_1}$, (1) reduces immediately to $\dfrac{y - y_1}{x - x_1} = m$ or

(2) $$y - y_1 = m(x - x_1).$$

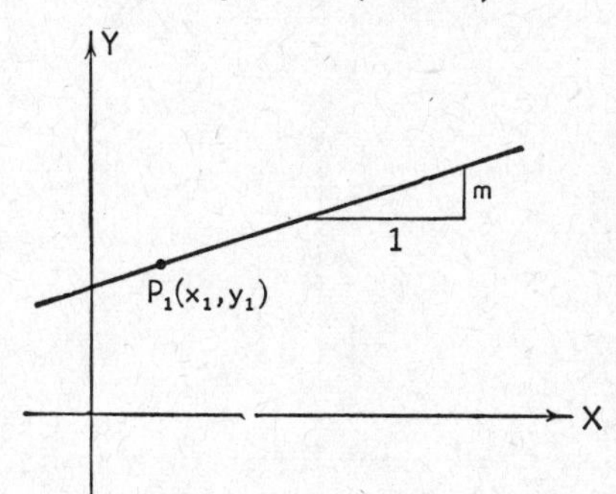

FIG. 43

III. *Slope-Intercept Form.* Specializing the point (x_1, y_1) in (2) to $(0, b)$ gives

(3) $$y = mx + b.$$

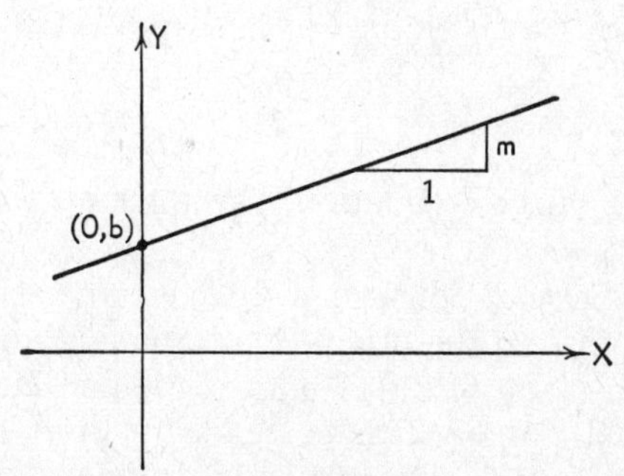

FIG. 44

Or again not all three of the coefficients A, B, C in the general equation are independent: some one of them is not zero and hence the equation could be divided by that one, leaving two effective *parameters* as they are sometimes called. For example if $B \neq 0$, the equation can be written $y = -\frac{A}{B}x - \frac{C}{B}$. Comparing this with (3) we note that in the general equation the slope of the line represented is given by $m = -A/B$, and the y-intercept by $-C/B$.

IV. *Intercept Form.* Specialization of the two points in (1) to the intercepts $(0, b)$ and $(a, 0)$ gives $(y - b)/x = -b/a$ or

$$\frac{x}{a} + \frac{y}{b} = 1. \tag{4}$$

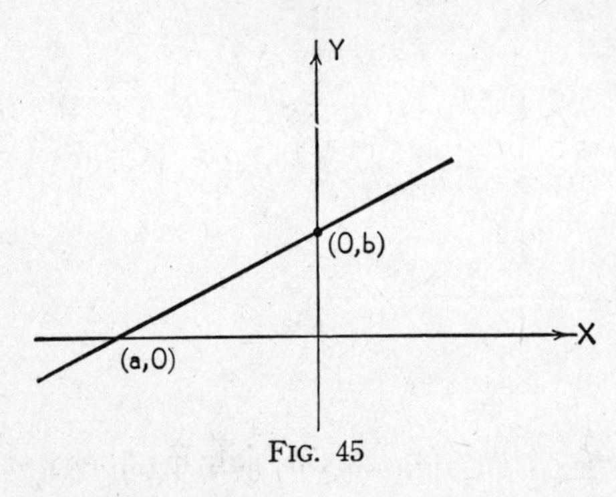

FIG. 45

Or again, dividing $Ax + By + C = 0$ by C gives, after some rearrangement,

$$\frac{x}{-\frac{C}{A}} + \frac{y}{-\frac{C}{B}} = 1,$$

whence, in the general equation, the intercepts are given by $-C/A$ and $-C/B$.

V. *Normal Form.* Consider a directed line segment OA of length p issuing from the origin O and making an angle θ with the positive direction of the X-axis. The line L which is perpendicular to OA and which passes through A is completely determined by the parameters p and θ. We wish to determine the equation of this (general) line L.

The coordinates of A are $A(p \cos \theta, p \sin \theta)$ and the slope of L is $-\cot \theta$ since L is perpendicular to OA which has slope $\tan \theta$. Hence, using the point-slope form (2) we obtain, as the *normal*

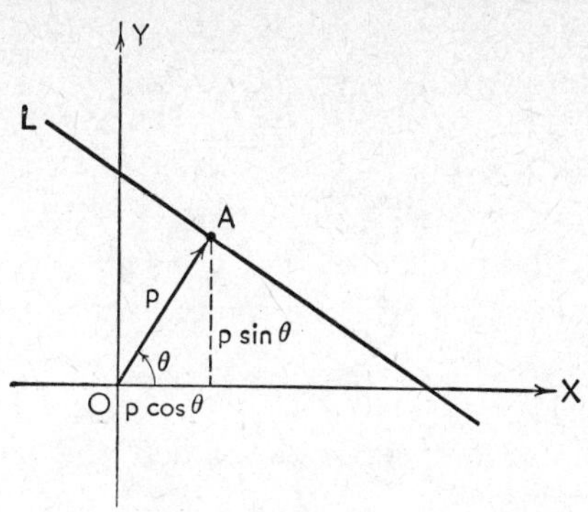

FIG. 46

form of the equation of L, $y - p \sin \theta = -\cot \theta(x - p \cos \theta)$ which reduces to

(5) $$x \cos \theta + y \sin \theta - p = 0.$$

This form of the equation of a straight line is called the *normal form* (sometimes *perpendicular form*) since its coefficients involve the parameters p and θ associated with the *normal* or *perpendicular* OA to the line. And it is important to note that

I. The coefficients of x and y are the direction cosines λ, μ of a *normal* to L. Equation (5) can be written

(5′) $$\lambda x + \mu y - p = 0.$$

II. The distance *from* the origin *to* the line is p.

Now comparing the two equations

$$\lambda x + \mu y - p = 0,$$
$$Ax + By + C = 0,$$

we see that the coefficients *A and B in the general equation of a line are direction numbers of any perpendicular line.* $A = k\lambda$, $B = k\mu$, $C = -kp$. To compute the proportionality factor k we square and add, thus

$$A^2 + B^2 = k^2(\lambda^2 + \mu^2)$$
$$= k^2$$

since $\lambda^2 + \mu^2 \equiv \cos^2 \theta + \sin^2 \theta \equiv 1$.

Hence

(6) $$k = \pm \sqrt{A^2 + B^2}.$$

We are now in a position to reduce the general equation to normal form by dividing through by (one of the values of) k.

$$(7) \quad \frac{A}{\pm\sqrt{A^2+B^2}}x + \frac{B}{\pm\sqrt{A^2+B^2}}y + \frac{C}{\pm\sqrt{A^2+B^2}} = 0.$$

Comparing (7) and (5) we see that the sign of $\sqrt{A^2+B^2}$ must be *opposite to that of* C so as to make the constant term read $-\left(\frac{C}{\pm\sqrt{A^2+B^2}}\right)$ where the parenthesis is p and therefore *positive.* If $C = 0$ we use for the radical the *same sign as that of* B.

Illustration 1. Write the equation of the line through $(2, 1)$, $(-6, 5)$. Reduce this to the general form.

Solution. Using the two-point form we get

$$\frac{y-1}{x-2} = \frac{5-1}{-6-2},$$

which reduces to $x + 2y - 4 = 0$.

Illustration 2. Find the equation of the line passing through $(3, -4)$ making an angle of $60°$ with the X-axis.

Solution. Point-slope form is called for and the equation is

$$y + 4 = \sqrt{3}(x - 3).$$

Illustration 3. Given the triangle $A(4, 5)$, $B(-2, 0)$, $C(2, -3)$. Find the equation of the median through C.

Solution. The midpoint of AB is $D(1, \frac{5}{2})$ and the equation of the desired median is

$$\frac{y-\frac{5}{2}}{x-1} = \frac{-3-\frac{5}{2}}{2-1},$$

or $11x + 2y - 16 = 0$.

Illustration 4. Find the intercepts of the line perpendicular to $2x + 3y - 7 = 0$ passing through $(1, 6)$.

Solution. The slope of $2x + 3y - 7 = 0$ is $-\frac{2}{3}$ and the slope of the desired line is $\frac{3}{2}$. Hence the equation is

$$y - 6 = \tfrac{3}{2}(x - 1).$$

When $y = 0$, $x = -3$ and when $x = 0$, $y = \frac{9}{2}$. The intercepts are therefore $(-3, 0)$ and $(0, \frac{9}{2})$.

Illustration 5. What are the direction cosines of a line perpendicular to $x - 5y + 3 = 0$?

Solution. The normal form of this equation is

$$-\frac{x}{\sqrt{26}} + \frac{5y}{\sqrt{26}} - \frac{3}{\sqrt{26}} = 0.$$

The direction cosines of a normal line are given by the coefficients of x and y and are therefore $\lambda = -\frac{1}{\sqrt{26}}$ and $\mu = \frac{5}{\sqrt{26}}$.

[The direction cosines of the line itself are $\lambda = \frac{5}{\sqrt{26}}$ and $\mu = \frac{1}{\sqrt{26}}$; see (2), p. 18.]

25. Distance from a Line to a Point. As a special application of the normal form of a line we show how very readily it yields the *distance from a line to a point.* Let $\lambda x + \mu y - p = 0$ be the normal form of some line L and let d be the distance from

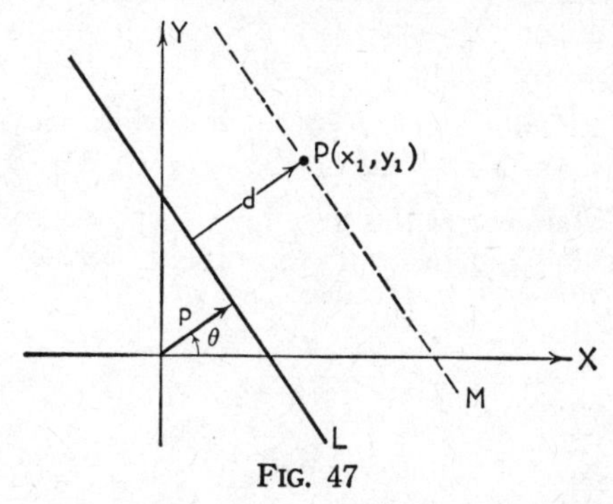

FIG. 47

this line to the point $P(x_1, y_1)$. Now (Fig. 47) $p + d$ could be considered a new "p," with the same θ, for the parallel line M through P. The normal form of M is

$$\lambda x + \mu y - (p + d) = 0. \tag{1}$$

As a by-product this tells us that any line parallel to L will have an equation of the form $\lambda x + \mu y - p' = 0$ where p' is a constant. Further, since (x_1, y_1) lies on (1), its coordinates satisfy (1) and this yields $\lambda x_1 + \mu y_1 - (p + d) = 0$, or, finally

$$d = \lambda x_1 + \mu y_1 - p. \tag{2}$$

This is the important formula for distance from a line to a point. It says that the function $\lambda x + \mu y - p$ (the left-hand member of the normal equation of a line) yields the distance from that line to the point P when the coordinates of P are substituted into it. This distance will be positive if the line separates the point P and the origin; it will be negative if P and the origin lie on the same side of the line.

Illustration 1. Find the distance from the line $x-y-5=0$ to the point (3, 4).

Solution. The normal form of the line is $\frac{x}{\sqrt{2}} - \frac{y}{\sqrt{2}} - \frac{5}{\sqrt{2}} = 0$ and the distance is $d = \frac{3}{\sqrt{2}} - \frac{4}{\sqrt{2}} - \frac{5}{\sqrt{2}} = -\frac{6}{\sqrt{2}}$. The distance is negative and the point lies on the same side of the line as the origin.

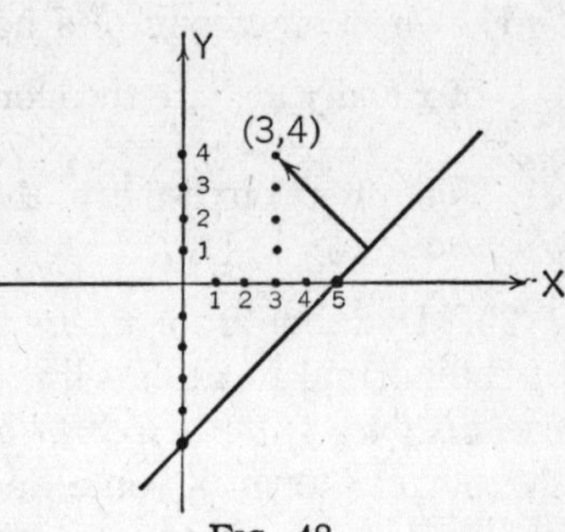

FIG. 48

Illustration 2. Find the locus of a point which moves so that it is always two units from the line $3x - 4y + 1 = 0$.

Solution. We interpret this to imply either of two lines parallel to the given line and two units (in absolute value) from it. The

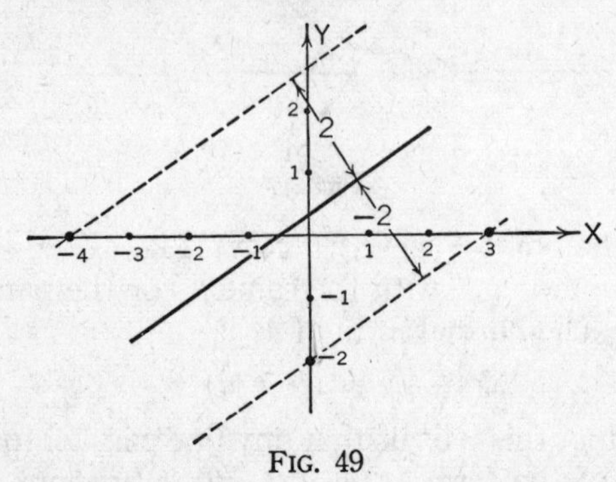

FIG. 49

normal form is $-\frac{3}{5}x + \frac{4}{5}y - \frac{1}{5} = 0$; and the equation of the desired locus is $-\frac{3}{5}x + \frac{4}{5}y - \frac{1}{5} \pm 2 = 0$. These two equations, one for either straight line, can be combined into the one equation $(3x - 4y - 9)(3x - 4y + 11) = 0$.

Illustration 3. Find the equations of the bisectors of the angles formed by $x + y + 2 = 0$ and $2x - 3y - 1 = 0$.

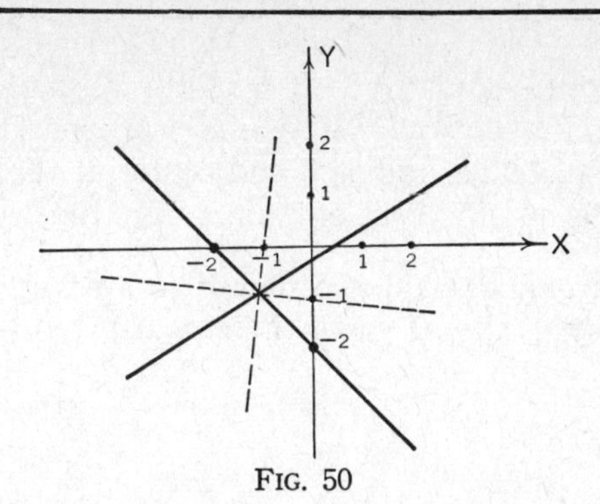

FIG. 50

Solution. A bisector is the locus of points equidistant from the two lines. The normal forms of the given lines are respectively

$$\frac{x + y + 2}{-\sqrt{2}} = 0 \quad \text{and} \quad \frac{2x - 3y - 1}{\sqrt{13}} = 0,$$

and if $P(X, Y)$ is (in absolute value) equidistant from the given lines, then

$$\frac{X + Y + 2}{-\sqrt{2}} = \pm \frac{2X - 3Y - 1}{\sqrt{13}}. \tag{1}$$

These are the equations of the angle bisectors. The capital letters X and Y are used to stress the fact that they are the coordinates of points on the desired locus whereas the small letters x and y refer to the coordinates of points on the given lines. We can of course shift now to lower-case letters. It is left as an exercise for the student to show that the two lines in (1) are perpendicular.

26. Systems of Lines. An important idea in mathematics is that of a *family* of lines or of curves. Since $x = k$ represents any line parallel to the Y-axis, we say that $x = k$ is the *family of lines* each parallel to the Y-axis. It is a *single-parameter family*, the parameter being k. Again $y = mx$ represents the family of lines through the origin with slope m; it also is a single-parameter family. The general equation $Ax + By + C = 0$, with *two effective* parameters, represents the family of all lines in the plane as does $\lambda x + \mu y - p = 0$, where the parameters are θ and p. The members of a family possess some common geometric property such as being parallel to a given line, passing through a given point, etc. Each member of the family $x - 2y + b = 0$ is parallel to the line $x - 2v - 3 = 0$; the members of the family $y - 2 = m(x + 5)$ pass through the point $(-5, 2)$; the family $\lambda x + \mu y - 5 = 0$ constitutes all lines tangent to the circle of radius 5 and center at the origin since each member is at a distance of 5 units from the origin.

27. Line through the Intersection of Two Given Lines. Let $L_1 \equiv a_1x + b_1y + c_1 = 0$ and $L_2 \equiv a_2x + b_2y + c_2 = 0$ be

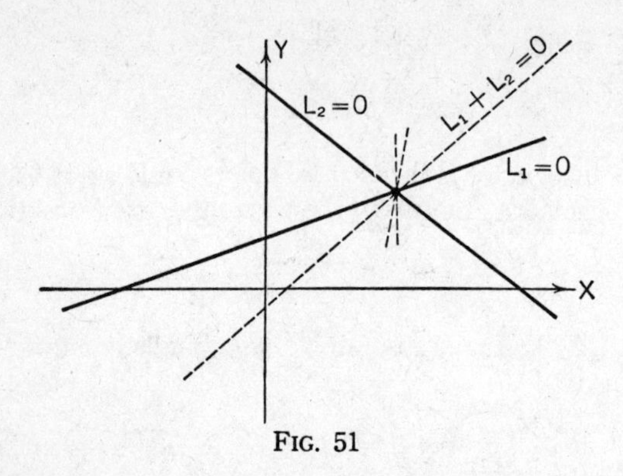

FIG. 51

two given lines and P their point of intersection. Consider the equation

$$(1) \quad L_1 + kL_2 \equiv (a_1x + b_1y + c_1) + k(a_2x + b_2y + c_2) = 0.$$

This is a line since it is of the first degree in x and y. But the coordinates of P will reduce each parenthesis in (1) to zero since, by hypothesis, P is the point of intersection, i.e., it lies on each line. Therefore P satisfies (1) and (1) represents the *family of lines through the intersection of* $L_1 = 0$ and $L_2 = 0$. It is not important which term k multiplies. As (1) stands it will not represent $L_2 = 0$; to do so $k(= 0)$ would have to multiply the first term.

Illustration 1. Find that member of the family of lines through the intersection of $x - y + 2 = 0$ and $2x + 3y - 5 = 0$ which passes through $(1, 5)$.

Solution. The family is given by $(x-y+2)+k(2x+3y-5)=0$; the line in question is a member of this family and passes through $(1, 5)$. Therefore

$$(1 - 5 + 2) + k(2 + 15 - 5) = 0,$$
$$-2 + 12\,k = 0,$$
$$k = \tfrac{1}{6}.$$

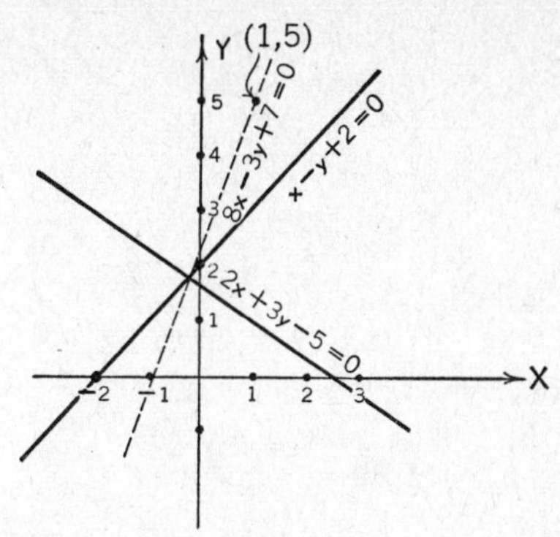

FIG. 52

The equation sought is

$$(x - y + 2) + \tfrac{1}{6}(2\,x + 3\,y - 5) = 0,$$

or, finally

$$8\,x - 3\,y + 7 = 0.$$

It is, in general, a waste of time to solve a problem of this type by first finding the point of intersection of the given lines.

Illustration 2. Find the equation of the line which passes through the point of intersection of $2\,x + y - 2 = 0$ and $x - y + 7 = 0$ and which is perpendicular to $x + 6\,y - 3 = 0$.

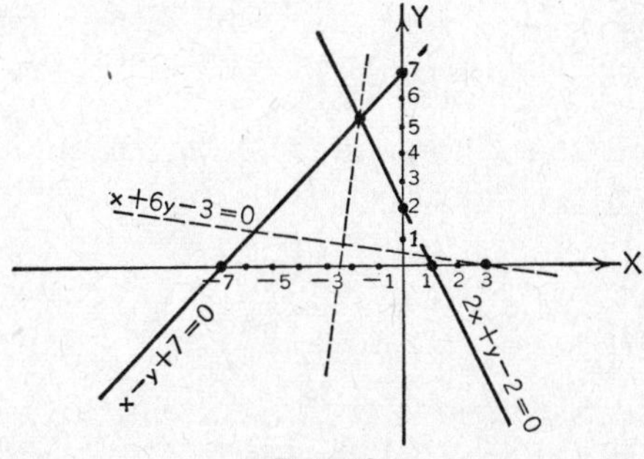

FIG. 53

Solution. The desired line is a member of the family

(1) $$(2\,x + y - 2) + k(x - y + 7) = 0$$

Its slope must be 6 since the slope of $x + 6y - 3 = 0$ is $-\frac{1}{6}$. Now the slope of each member of the family is of the form $-\dfrac{2+k}{1-k}$ which is gotten by solving (1) for y and then picking out the coefficient of x. (See the slope-intercept form $y = mx + b$.) Therefore $6 = -\dfrac{2+k}{1-k}$ or $k = \dfrac{8}{5}$. Hence the line sought has the equation

$$(2x + y - 2) + \tfrac{8}{5}(x - y + 7) = 0,$$

which reduces to

$$18x - 3y + 46 = 0.$$

The student should check this, for practice, by solving the given equations for the point of intersection $(-\frac{5}{3}, \frac{16}{3})$ and by applying the point-slope form of the straight line.

Illustration 3. What is the slope of the line joining the origin with the point of intersection of $x - 4y + 1 = 0$ and $3x + y + 2 = 0$?

Solution. The line is a member of the family

$$(x - 4y + 1) + k(3x + y + 2) = 0$$

and passes through (0, 0). Therefore $1 + 2k = 0$ or $k = -\frac{1}{2}$, and the equation of the line becomes

$$(x - 4y + 1) - \tfrac{1}{2}(3x + y + 2) = 0,$$
$$x + 9y = 0.$$

The slope of this line is $-\frac{1}{9}$.

In passing we note that if the original lines are parallel, $ax + by + c = 0$ and $ax + by + c' = 0$, then every member of the family $(ax + by + c) + k(ax + by + c') = 0$ will be parallel to them. (The point of intersection is at infinity.)

28. Condition That Three Lines Be Concurrent. In algebra it is proved that three linear equations in two unknowns (lines),

$$a_1x + b_1y + c_1 = 0,$$
$$a_2x + b_2y + c_2 = 0,$$
$$a_3x + b_3y + c_3 = 0,$$

will have a common solution (be concurrent, meet in a point), only when the determinant of the coefficients is zero. This is

$$\begin{vmatrix} a_1 & b_1 & c_1 \\ a_2 & b_2 & c_2 \\ a_3 & b_3 & c_3 \end{vmatrix} = 0. \tag{1}$$

This condition is therefore necessary; it is also sufficient provided the slopes of the lines are distinct. (The case of parallel or coinciding lines is a little more involved.)

Illustration 1. Show that the three lines $x-y+6=0$, $2x+y-5=0$, $-x-2y+11=0$ are concurrent.

Solution. Since the slopes are distinct and since

$$\begin{vmatrix} 1 & -1 & 6 \\ 2 & 1 & -5 \\ -1 & -2 & 11 \end{vmatrix} = 11 - 24 - 5 + 6 + 22 - 10 = 0$$

the lines meet in a common point.

The student may show that the point of intersection of the first two lines is $(-\frac{1}{3}, \frac{17}{3})$ and that this point lies on the third line.

Illustration 2. Find k so that $x+y+1=0$, $kx-y+3=0$, and $4x-5y+k=0$ will be concurrent.

Solution. We must have

$$\begin{vmatrix} 1 & 1 & 1 \\ k & -1 & 3 \\ 4 & -5 & k \end{vmatrix} = 31 - 6k - k^2 = 0.$$

Since this is a quadratic equation there will be two values of k that will make the lines concurrent, namely $k = -3 \pm 2\sqrt{10}$, because in either case the slopes will be distinct.

29. Condition That Three Points Be Collinear. The two-point form of the equation of a line is

$$\frac{y-y_1}{x-x_1} = \frac{y_2-y_1}{x_2-x_1},$$

which reduces to

$$xy_1 + x_1y_2 + x_2y - x_2y_1 - x_1y - xy_2 = 0$$

upon simplification. And this can be written as a determinantal equation

$$\begin{vmatrix} x & y & 1 \\ x_1 & y_1 & 1 \\ x_2 & y_2 & 1 \end{vmatrix} = 0 \tag{1}$$

to give another two-point form of the equation of a straight line.

We can readily check that (1) is the equation of the line through (x_1, y_1), (x_2, y_2) because when either set is substituted into (1) for x, y the determinant obviously vanishes since then two rows are alike.

Now consider a third point (x_3, y_3). It will lie on line (1) if and only if

$$\begin{vmatrix} x_1 & y_1 & 1 \\ x_2 & y_2 & 1 \\ x_3 & y_3 & 1 \end{vmatrix} = 0. \tag{2}$$

(We may interchange the 1st and 3rd rows without changing anything in this equation.) This is the necessary and sufficient condition that three points be collinear (lie on a line).

Illustration 1. Show that the three points (1, 2), (7, 6), (4, 4) are collinear.

Solution. Now

$$\begin{vmatrix} 1 & 2 & 1 \\ 7 & 6 & 1 \\ 4 & 4 & 1 \end{vmatrix} = 6 + 28 + 8 - 24 - 14 - 4 = 0.$$

Therefore the points lie on a line.

Illustration 2. Find the value of k so that $(1, -3)$, $(-2, 5)$. $(4, k)$ lie on a line.

Solution. We must have

$$\begin{vmatrix} 1 & -3 & 1 \\ -2 & 5 & 1 \\ 4 & k & 1 \end{vmatrix} = -3\,k - 33 = 0.$$

Therefore $k = -11$.

30. Résumé of Straight Line Formulae.

	EQUATION	FORM
(1)	$\dfrac{y - y_1}{x - x_1} = \dfrac{y_2 - y_1}{x_2 - x_1}$	Two-point
(2)	$y - y_1 = m(x - x_1)$	Point-slope
(3)	$y = mx + b$	Slope-intercept
(4)	$\dfrac{x}{a} + \dfrac{y}{b} = 1$	Intercept
(5)	$\lambda x + \mu y - p = 0$, or $x \cos \theta + y \sin \theta - p = 0$	Normal
(6)	$y = mx$	Through origin
(7)	$x = k$	Perpendicular to X-axis
(8)	$y = k$	Perpendicular to Y-axis

(9) $Ax + By + C = 0$ General

(10) $\begin{vmatrix} x & y & 1 \\ x_1 & y_1 & 1 \\ x_2 & y_2 & 1 \end{vmatrix} = 0$ Determinant

EXERCISES

Find the equation of each of the following lines.

1. With intercepts $(1, 0)$ and $(0, -3)$. *Ans.* $3x - y - 3 = 0$.

2. Through the x-intercept of $2x + 3y - 8 = 0$ and perpendicular to the line joining $(1, 3)$, $(2, -1)$. *Ans.* $x - 4y - 4 = 0$.

3. Through the intersection of the diagonals of the trapezoid $A(1, 0)$, $B(0, 2)$, $C(-3, -1)$, $D(-1, -5)$ and parallel to the Y-axis.
Ans. $3x + 1 = 0$.

4. With slope 2 and tangent to the circle of radius 5 with center at the origin. *Ans.* $2x - y \pm 5\sqrt{5} = 0$.

5. Through $(1, 2)$ and the median point of the triangle $(2, 0)$, $(-5, 1)$, $(-3, 7)$. *Ans.* $2x + 9y - 20 = 0$.

6. Parallel to and $+3$ units from the perpendicular bisector of the line segment $(1, -2)$, $(-3, 8)$. *Ans.* $2x - 5y + (17 + 3\sqrt{29}) = 0$.

Find the equation of each of the following families of lines.

7. Perpendicular to $x - 3y + 6 = 0$. *Ans.* $3x + y + c = 0$.

8. Through the intersection of $x + y - 5 = 0$ and $4x + y + 1 = 0$.
Ans. $(x + y - 5) + k(4x + y + 1) = 0$.

9. Through the center of the rectangle formed by the lines $x = a$, $x = b$, $y = c$, $y = d$. *Ans.* $y - \frac{c + d}{2} = m\left(x - \frac{a + b}{2}\right)$.

10. Through the origin and intersecting the finite line segment joining $(2, 4)$ and $(6, 1)$. *Ans.* $y = mx$, $\frac{1}{6} \leqq m \leqq 2$.

Solve the following problems.

11. Show that $3x - 5y + 8 = 0$, $x + 2y - 4 = 0$, and $4x - 3y + 4 = 0$ are concurrent.

12. Find k so that $kx + y + 1 = 0$, $x + ky + 1 = 0$, and $x + y + k = 0$ are concurrent. *Ans.* $k = 1, 1, -2$.

13. Show that the three points $(1, 9)$, $(-2, 3)$, $(-5, -3)$ are collinear.

CHAPTER V

THE CIRCLE

31. Introduction. In the previous chapter we examined the linear equation

(1) $$Ax + By + C = 0. \quad \text{(Straight line)}$$

In this and the next few chapters we study the *second-degree* equation

(2) $$Ax^2 + Bxy + Cy^2 + Dx + Ey + F = 0. \quad \text{(Conic section)}$$

We shall take up special forms of this equation, discussing in detail the locus in each case.

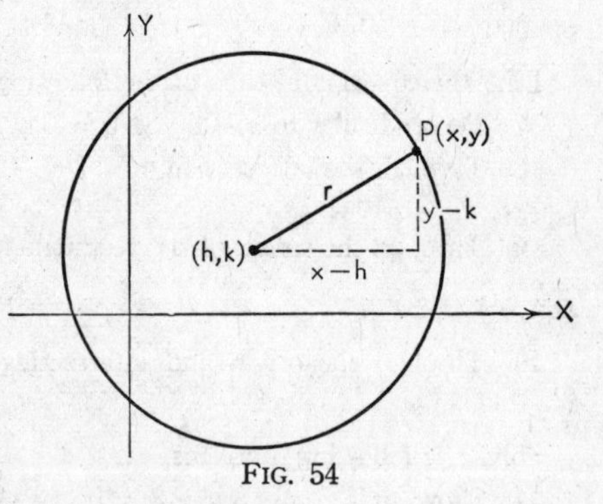

FIG. 54

32. Standard Form of the Equation of a Circle. Let $P(x, y)$ be a point in the plane which moves so that it is always a constant distance r from the fixed point (h, k). The locus of P is obviously a circle of radius r and center at (h, k); to find it we need only apply the distance formula. Thus

$$\sqrt{(x - h)^2 + (y - k)^2} = r,$$

(1) $$(x - h)^2 + (y - k)^2 = r^2$$

is the equation in *standard form.* In this form the equation clearly exhibits the center and radius. We may reduce (1) to the form

(2) $$x^2 + y^2 - 2hx - 2ky + (h^2 + k^2 - r^2) = 0.$$

In this form the equation appears as a special case of the general second-degree equation (2) of § 31 above: the coefficients of x^2 and y^2 are the same ($A = C$) and there is no xy-term ($B = 0$).

Conversely any equation of the form $x^2+y^2+ax+by+c=0$ can be reduced to form (1) and hence represents a circle. Therefore *a necessary and sufficient condition that* $Ax^2 + Bxy + Cy^2 + Dx + Ey + F = 0$ *represent a circle is that* $A = C$ *and* $B = 0$. It is not necessary that $A = C = 1$ as in (2) since the coefficient of x^2 and y^2 is not zero (for in that case the equation would be linear and represent a line) and hence could be divided out, reducing it to unity.

If the center is at the origin (1) reduces to

$$x^2 + y^2 = r^2. \tag{3}$$

33. Reduction of the General Equation to Standard Form. The equation

$$x^2 + y^2 + ax + by + c = 0 \tag{1}$$

is often called the *general equation* of the circle although it is no more general than (1) § 32. It is readily reduced to standard form. Complete the square first on the x^2- and x-terms, then on the y^2- and y-terms. (To complete the square on $z^2 + Az$ add, and subtract, *the square of half the coefficient of* z.) We get

$$x^2 + ax + \frac{a^2}{4} + y^2 + by + \frac{b^2}{4} + c - \frac{a^2}{4} - \frac{b^2}{4} = 0,$$

or

$$\left(x + \frac{a}{2}\right)^2 + \left(y + \frac{b}{2}\right)^2 = \frac{a^2 + b^2 - 4c}{4}. \tag{2}$$

This is the equation of a circle with center at $(-a/2, -b/2)$ and radius $r = \frac{1}{2}\sqrt{a^2 + b^2 - 4c}$. Note that whereas $Ax + By + C = 0$, with *real* coefficients, always plots a (real) line, the circle (2) will be a real circle only if $a^2 + b^2 - 4c > 0$. If $a^2 + b^2 - 4c = 0$ equation (2) represents one point only (circle of zero radius).

Illustration 1. Find the equation of the circle with center at $(-2, 3)$ and radius 6.

Solution. Directly from (1) § 32 we have as the equation

$$(x + 2)^2 + (y - 3)^2 = 36.$$

Illustration 2. Find the center and radius of the circle

$$2x^2 + 2y^2 - 5x + 4y - 7 = 0.$$

Solution. The given circle has the equation

$$x^2 + y^2 - \tfrac{5}{2}x + 2y = \tfrac{7}{2}.$$

Completing the square we get

$$x^2 - \tfrac{5}{2}x + \tfrac{25}{16} + y^2 + 2y + 1 = \tfrac{7}{2} + \tfrac{25}{16} + 1,$$

or

$$(x - \tfrac{5}{4})^2 + (y + 1)^2 = \tfrac{97}{16}.$$

The center is at $(\tfrac{5}{4}, -1)$; the radius is $\tfrac{1}{4}\sqrt{97}$.

34. Circle Determined by Three Conditions. Since the equation of a circle has three effective parameters (h, k, r or a, b, c), in general some three conditions can be imposed upon them which will determine a circle, unique or otherwise.

I. *Circles through Points.* We can find the equation of the circle through three points $P_1(x_1, y_1)$, $P_2(x_2, y_2)$, $P_3(x_3, y_3)$ as follows. Take the equation in general form

$$x^2 + y^2 + ax + by + c = 0.$$

Substituting the coordinates of the three points in this equation we get the three linear equations in the three unknowns a, b, c,

$$(1) \qquad \begin{aligned} ax_1 + by_1 + c &= -x_1^2 - y_1^2, \\ ax_2 + by_2 + c &= -x_2^2 - y_2^2, \\ ax_3 + by_3 + c &= -x_3^2 - y_3^2. \end{aligned}$$

The simultaneous solution of these three equations for a, b, and c will yield the coefficients in the equation of the circle.

Illustration 1. Find the equation of the circle passing through (1, 2), (−2, 0), (−1, −5).

Solution. The simultaneous equations are

$$\begin{aligned} a + 2b + c &= -5, \\ -2a \qquad + c &= -4, \\ -a - 5b + c &= -26. \end{aligned}$$

The determinant of the coefficients is $D = \begin{vmatrix} 1 & 2 & 1 \\ -2 & 0 & 1 \\ -1 & -5 & 1 \end{vmatrix} = 17.$

$$a = \frac{\begin{vmatrix} -5 & 2 & 1 \\ -4 & 0 & 1 \\ -26 & -5 & 1 \end{vmatrix}}{17} = -\frac{49}{17}, \quad b = \frac{\begin{vmatrix} 1 & -5 & 1 \\ -2 & -4 & 1 \\ -1 & -26 & 1 \end{vmatrix}}{17} = \frac{65}{17},$$

$$c = \frac{\begin{vmatrix} 1 & 2 & -5 \\ -2 & 0 & -4 \\ -1 & -5 & -26 \end{vmatrix}}{17} = -\frac{166}{17}.$$

The equation of the circle is therefore

$$17\,x^2 + 17\,y^2 - 49\,x + 65\,y - 166 = 0.$$

Note that if D were zero it would be impossible to solve for a, b, c. But $D = 0$ is the condition that the three points be collinear [see (2) § 29] in which case no circle would exist.

II. *Circles Tangent to Lines.* Instead of specifying that the circle pass through certain points we may require that it be tangent to certain lines or that its center lie on a given line. Combinations of point and line conditions may be used to determine a circle (or circles).

Illustration 2. The three lines $x=0,\ y=0,\ 3\,x+4\,y-1=0$ form a triangle. Find the equation of the inscribed circle.

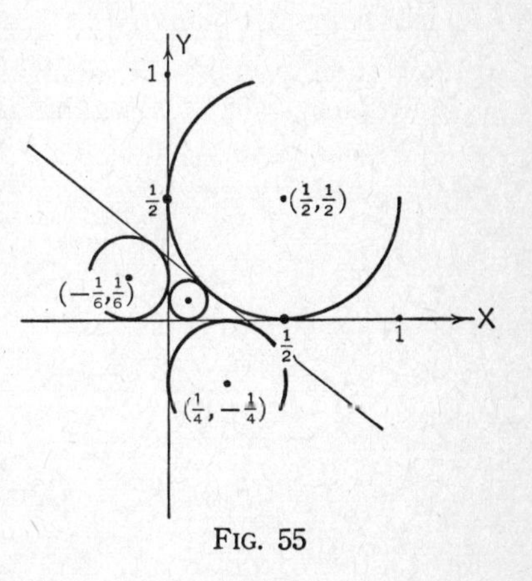

FIG. 55

Solution. Let the equation of the circle be

$$(x - h)^2 + (y - k)^2 = r^2.$$

The center is equidistant from the sides of the triangle; hence

$$h = k = r. \tag{1}$$

The normal form of $3\,x + 4\,y - 1 = 0$ is $\dfrac{3\,x + 4\,y - 1}{5} = 0.$

Since the left-hand member of this equation gives the distance *from* the line *to* a point and since the point in question — the center of the circle — and the origin are on the same side of the line, this particular distance is negative and we must equate

$$-r = \frac{3\,h + 4\,k - 1}{5}. \tag{2}$$

Solving (1) and (2) simultaneously we get $h = k = r = \frac{1}{12}$. The inscribed circle has the equation

$$(x - \tfrac{1}{12})^2 + (y - \tfrac{1}{12})^2 = \tfrac{1}{144}.$$

In a similar manner we may determine the circles externally tangent to this triangle. For the one in the I, II, and IV quadrants respectively we solve simultaneously $h = k = r$ and $5r = 3h + 4k - 1$, $-h = k = r$ and $-5r = 3h + 4k - 1$, $h = -k = r$ and $-5r = 3h + 4k - 1$.

Illustration 3. Find the equation of the circle tangent to the two axes and passing through the point $(1, -7)$.

Solution. The conditions of the problem lead to the simultaneous equations

$$(1) \qquad (1 - h)^2 + (-7 - k)^2 = r^2,$$
$$(2) \qquad h = -k = r,$$

which yield $h = 8 \pm \sqrt{14}$, $k = -8 \mp \sqrt{14}$, $r = 8 \pm \sqrt{14}$. From these the equations can be written down. There are two circles satisfying the conditions.

Illustration 4. Find the equation of the circle which is tangent to the Y-axis, which passes through the point $(-1, -1)$, and the center of which is on the line $2x + y + 4 = 0$.

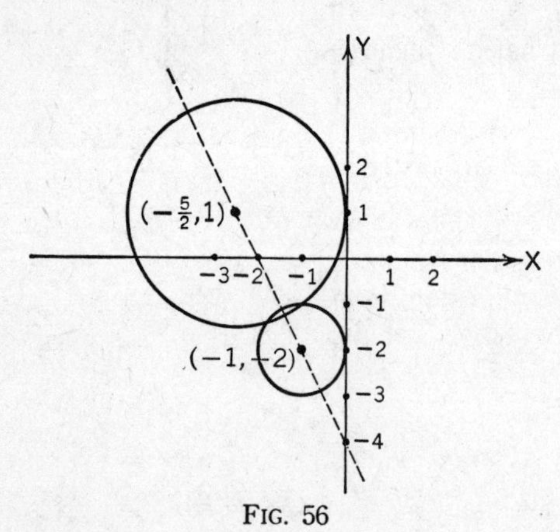

FIG. 56

Solution. The simultaneous equations are

$$(1) \qquad -h = r,$$
$$(2) \qquad (-1 - h)^2 + (-1 - k)^2 = r^2,$$
$$(3) \qquad 2h + k + 4 = 0.$$

These have the two sets of solutions $h = -1$, $k = -2$, $r = 1$ and $h = -\frac{5}{2}$, $k = 1$, $r = \frac{5}{2}$. Again there are two circles and their equations can now be written down.

EXERCISES

1. Write the equation of the circle with center at $(1, -3)$ and radius 4.
Ans. $(x-1)^2 + (y+3)^2 = 16$.

2. Find the center and radius of the circle $x^2 + y^2 - 4x + y - 1 = 0$.
Ans. $(2, -\frac{1}{2})$, $r = \frac{1}{2}\sqrt{21}$.

3. Show that the equation $x^2 + y^2 + 2x - 4y + 8 = 0$ plots no real point.

4. Prove that the locus of a point, the sum of whose squares of distances from two fixed points (x_1, y_1) and (x_2, y_2) is a constant. is a circle real or imaginary.

Find the equations of the following circles.

5. Through $(1, 2)$, $(0, -1)$, $(-1, 1)$. *Ans.* $x^2 + y^2 - x - y - 2 = 0$.

6. Through $(-1, 2)$ and tangent to the axes.
Ans. $(x+1)^2 + (y-1)^2 = 1$ and $(x+5)^2 + (y-5)^2 = 25$.

7. Through the point of intersection of $2x - y + 7 = 0$ and $3x + y + 8 = 0$ with center at the origin. *Ans.* $x^2 + y^2 = 10$.

8. Through $(2, -1)$ and $(-2, 0)$ with center on $2x - y - 1 = 0$.
Ans. $(x + \frac{1}{4})^2 + (y + \frac{3}{2})^2 = \frac{85}{16}$.

35. Equation of a Line Tangent to a Circle.

I. *At a Given Point.* Let the circle be $(x-h)^2 + (y-k)^2 = r^2$ and $P(x_1, y_1)$ a point on it. Since the slope of the line joining the center (h, k) and P is $(y_1 - k)/(x_1 - h)$, the slope of the tangent will be $-(x_1 - h)/(y_1 - k)$ and the equation of the line tangent to the circle at point P becomes (point-slope form)

$$y - y_1 = -\frac{x_1 - h}{y_1 - k}(x - x_1).$$

Making use of the relation $(x_1 - h)^2 + (y_1 - k)^2 = r^2$ the equation of the tangent reduces to

$$(1) \qquad (x - h)(x_1 - h) + (y - k)(y_1 - k) = r^2.$$

Because of its symmetry this is easily remembered.

If the circle is centered at the origin, $x^2 + y^2 = r^2$, the equation of the tangent at P is

$$(2) \qquad xx_1 + yy_1 = r^2.$$

The tangent to $x^2 + y^2 + ax + by + c = 0$ at P is

$$(3) \qquad xx_1 + yy_1 + \frac{a}{2}(x + x_1) + \frac{b}{2}(y + y_1) + c = 0.$$

II. *From a Point outside the Circle.* If the point $P'(x', y')$ is outside the circle there will be two tangents from it to the circle. For the circle $x^2 + y^2 = r^2$ their equations will be

$$(4) \qquad y - y' = \frac{x'y' \pm r\sqrt{x'^2 + y'^2 - r^2}}{x'^2 - r^2}(x - x'),\ x'^2 \neq r^2$$

III. *With a Given Slope.* The parallel tangents, with slope m, to $(x - h)^2 + (y - k)^2 = r^2$ are

$$(5) \qquad y - k = m(x - h) \pm r\sqrt{1 + m^2},$$

which reduce to

$$(6) \qquad y = mx \pm r\sqrt{1 + m^2}$$

for the circle $x^2 + y^2 = r^2$.

Illustration 1. Find the equation of the tangent to each circle at the point indicated.

(a) $(x - 1)^2 + (y + 2)^2 = 9, \quad (1, 1);$

(b) $x^2 + y^2 + 4x - 5y + 9 = 0, \quad (-1, 3);$

(c) $x^2 + y^2 = 3, \quad (1, -\sqrt{2}).$

Solution. The equations of the tangents are, upon direct substitution in (1), (3), and (2) respectively,

(a) $(x - 1)(1 - 1) + (y + 2)(1 + 2) = 9,$

or $y = 1;$

(b) $x(-1) + y(3) + 2(x - 1) - \frac{5}{2}(y + 3) + 9 = 0,$

or $2x + y - 1 = 0;$

(c) $x - \sqrt{2}\,y = 3.$

Illustration 2. Find the equations of the tangents to the circle $x^2 + y^2 = 16$ from the point $(-3, 7)$.

Solution. From (4) we have, as the desired equations,

$$y - 7 = \frac{-21 \pm 4\sqrt{9 + 49 - 16}}{9 - 16}(x + 3),$$

or

$$y = 10 \pm \tfrac{4}{7}\sqrt{42}\,(x + 3).$$

Illustration 3. Find the equations of the tangents to $(x + 2)^2 + (y - 1)^2 = 9$ with slope -1.

Solution. Equation (5) yields

$$y - 1 = -(x + 2) \pm 3\sqrt{2},$$
$$\text{or } x + y + 1 \pm 3\sqrt{2} = 0.$$

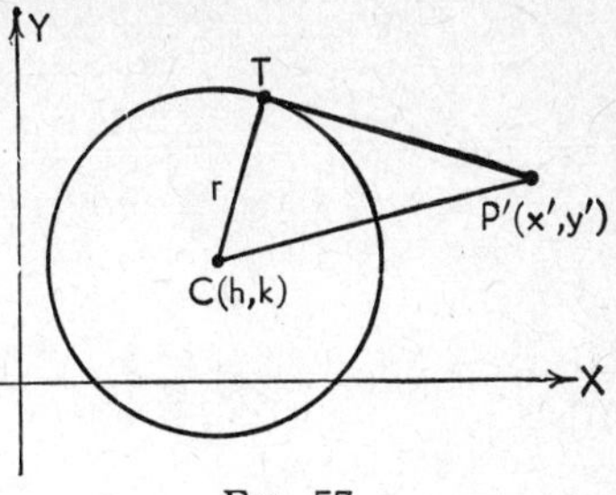

FIG. 57

36. Length of a Tangent. Let $P'(x', y')$ be a point outside of $(x-h)^2+(y-k)^2-r^2=0$. Draw $P'T$, TC, and CP' (Fig. 57) where T is the point of contact of the tangent from P' and C the center of the circle.

Now $\overline{P'T}^2 = \overline{P'C}^2 - \overline{CT}^2.$

But $\overline{P'C}^2 = (x' - h)^2 + (y' - k)^2$

and $\overline{CT}^2 = r^2.$

Hence $\overline{P'T}^2 = (x' - h)^2 + (y' - k)^2 - r^2$

gives the square of the length of a tangent.

Illustration. Find the length of a tangent to the circle $x^2 + y^2 - 6x + 2y - 6 = 0$ from $P'(-2, 0)$.

Solution. We first must reduce the equation of the circle to standard form. This is $(x - 3)^2 + (y + 1)^2 = 16$. The length of a tangent is therefore

$$P'T = \sqrt{25 + 1 - 16} = \sqrt{10}.$$

37. Systems of Circles. Many of the ideas of systems, or families, of straight lines (§ 26) carry over to circles. The family of concentric circles with center at the origin has the equation $x^2 + y^2 = r^2$; the family of circles tangent to the axes has the equation $(x - h)^2 + (y - h)^2 = h^2$; the equation $(x - h)^2 + (y - 1)^2 = 25$ represents the family of circles with centers on the line $y = 1$ and radius 5. All of these are single-parameter families. The family of circles passing through the origin has the equation $x^2 + y^2 + ax + by = 0$ and is an example of a two-parameter family. The general equation of all circles in the plane has three parameters.

Let $P_1(x_1, y_1)$ and $P_2(x_2, y_2)$ be the two points of intersection of the two circles $S_1 \equiv x^2 + y^2 + a_1x + b_1y + c_1 = 0$ and $S_2 \equiv x^2 + y^2 + a_2x + b_2y + c_2 = 0$. Consider the equation

$$(1)\quad S_1 + kS_2 \equiv (x^2 + y^2 + a_1x + b_1y + c_1) + k(x^2 + y^2 + a_2x + b_2y + c_2) = 0.$$

This represents a circle since it is of the second degree, the coefficients of x^2 and y^2 are the same, namely $(1 + k)$, and there

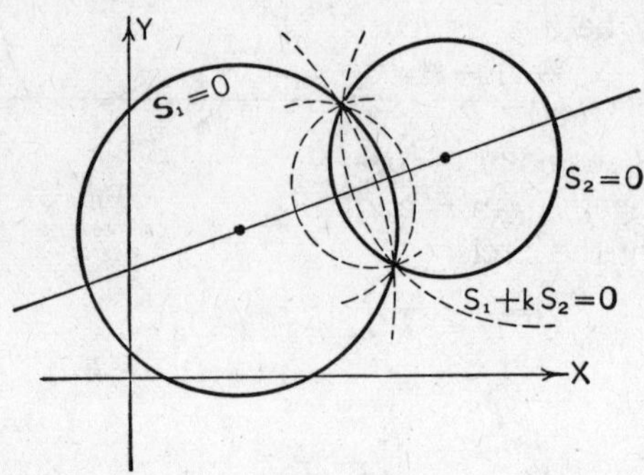

FIG. 58

is no xy-term. Moreover P_1 and P_2 both are points on it since the coordinates of either reduce each parenthesis of (1), and hence the whole equation, to zero. Therefore (1) represents the *family of circles through the points of intersection of the two given circles.* A particular member of this family may be determined by specifying that it satisfy some further condition.

Illustration. Find that member of the family of circles through the intersections of $x^2 + y^2 - 5x + y - 4 = 0$ and $x^2 + y^2 + 2x - 3y - 1 = 0$ which passes through $(1, -5)$.

Solution. The family is given by

$$(x^2 + y^2 - 5x + y - 4) + k(x^2 + y^2 + 2x - 3y - 1) = 0,$$

and the particular member passes through $(1, -5)$. Therefore

$$(1 + 25 - 5 - 5 - 4) + k(1 + 25 + 2 + 15 - 1) = 0,$$
$$k = -\tfrac{2}{7}.$$

The equation of the desired circle is

$$5x^2 + 5y^2 - 39x + 13y - 26 = 0.$$

38. Radical Axis. A very special and interesting case arises when $k = -1$, for then the system $S_1 + kS_2$ reduces to

$$(1)\quad S_1 - S_2 \equiv (a_1 - a_2)x + (b_1 - b_2)y + (c_1 - c_2) = 0.$$

This is not a circle at all. It is instead a straight line and is called the *radical axis* of the two circles $S_1 = 0$ and $S_2 = 0$ (R_1R_2 in Fig. 59).

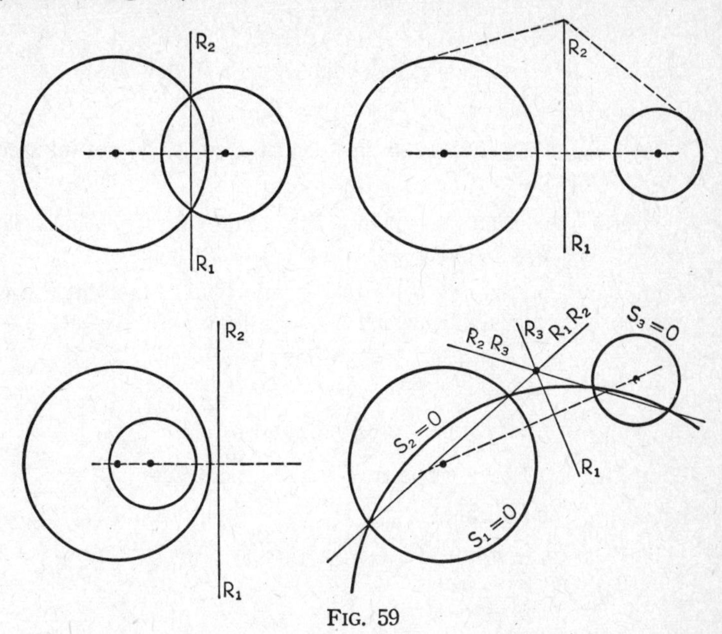

FIG. 59

The radical axis possesses the following properties:

I. It is the line of the common chord if the circles intersect in distinct real points;

II. It is the common tangent line if the circles intersect in coincident points (are tangent internally or externally);

III. It is a real straight line even though the circles do not intersect in real points;

IV. It is the locus of points from which tangents of equal length can be drawn to the two circles;

V. It is perpendicular to the line of centers of the two circles;

VI. It does not exist (is at infinity) if the defining circles are concentric;

VII. The radical axes of three circles, taken in pairs, intersect in a point called the *radical center*. (When the centers of the circles are collinear, the radical axes will be parallel lines and the radical center will be at infinity.)

Illustration 1. Find the equation of the common chord of the two circles $x^2 + y^2 + 3x - 2y - 7 = 0$ and $x^2 + y^2 - x - y + 2 = 0$.

Solution. The equation is immediately written down as

$$4x - y - 9 = 0.$$

It is left for the student to show that the circles do not intersect in real points.

Illustration 2. Find the points of intersection of the circles $x^2 + y^2 - 25 = 0$ and $x^2 + y^2 + x + y - 20 = 0$.

Solution. It is easier to solve the equation of the first circle with that of the radical axis, which is $x + y + 5 = 0$. We get

$$x^2 + (-x - 5)^2 - 25 = 0,$$

or

$$x^2 + 5x = 0.$$

The points of intersection are $(0, -5)$ and $(-5, 0)$.

Illustration 3. Prove property V in the general case.

Solution. The radical axis is

$$(1) \qquad (a_1 - a_2)x + (b_1 - b_2)y + (c_1 - c_2) = 0$$

with slope $-\dfrac{a_1 - a_2}{b_1 - b_2}$.

The slope of the line of centers will be, since the centers themselves have coordinates $\left(-\dfrac{a_1}{2}, -\dfrac{b_1}{2}\right)$ and $\left(-\dfrac{a_2}{2}, -\dfrac{b_2}{2}\right)$,

$$\frac{-\dfrac{b_2}{2} + \dfrac{b_1}{2}}{-\dfrac{a_2}{2} + \dfrac{a_1}{2}} = \frac{b_1 - b_2}{a_1 - a_2}.$$

This is the negative reciprocal of the slope of the radical axis and hence the property is demonstrated.

39. Orthogonal Circles. Two circles will be orthogonal (intersect at right angles) if their tangents at a point of intersection are at right angles. Hence the tangent to one circle passes through the center of the other and vice versa. We wish to determine the condition on the coefficients that two circles shall be orthogonal.

Let the equations of the circles be $x^2+y^2+a_1x+b_1y+c_1=0$, and $x^2 + y^2 + a_2x + b_2y + c_2 = 0$. Now (Fig. 60)

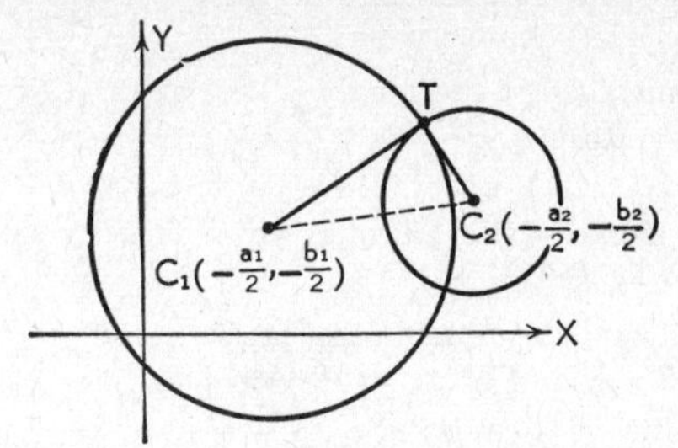

FIG. 60

$$\overline{C_1C_2}^2 = \overline{C_1T}^2 + \overline{C_2T}^2.$$

But
$$\overline{C_1C_2}^2 = \left(\frac{a_1}{2} - \frac{a_2}{2}\right)^2 + \left(\frac{b_1}{2} - \frac{b_2}{2}\right)^2,$$

and, by (2) § 33,
$$\overline{C_1T}^2 = r_1^2 = \frac{a_1^2 + b_1^2 - 4\,c_1}{4},$$

$$\overline{C_2T}^2 = r_2^2 = \frac{a_2^2 + b_2^2 - 4\,c_2}{4}.$$

Hence

$$\left(\frac{a_1}{2} - \frac{a_2}{2}\right)^2 + \left(\frac{b_1}{2} - \frac{b_2}{2}\right)^2 = \frac{a_1^2 + b_1^2 - 4\,c_1 + a_2^2 + b_2^2 - 4\,c_2}{4},$$

which reduces to the simple relation

(1) $$a_1a_2 + b_1b_2 = 2(c_1 + c_2),$$

which is the condition (necessary and sufficient) that the two circles be orthogonal.

Illustration. Show that the two circles $x^2 + y^2 - 3\,x + 2\,y - 3 = 0$ and $x^2 + y^2 + 2\,x + y + 1 = 0$ are orthogonal.

Solution. The condition (1) for orthogonality reduces to

$$(-3)(2) + (2)(1) = 2(-3 + 1).$$

Hence the circles intersect at right angles.

EXERCISES

1. Find the equation of the tangent to the circle $(x - 1)^2 + (y + \frac{1}{2})^2 = 5$ at the point $(3, \frac{1}{2})$. *Ans.* $4\,x + 2\,y - 13 = 0$.

2. Find the equations of the tangents to $x^2 + y^2 = 4$ from (1, 6). *Ans.* $3\,y + 6\,x - 24 = \pm\, 2\sqrt{33}(x - 1)$.

3. Find the tangents to $(x + 1)^2 + (y - 5)^2 = 9$ with slope -2.
Ans. $2x + y - 3 \pm 3\sqrt{5} = 0$.

4. Find the length of a tangent to the circle $(x - 2)^2 + (y - 4)^2 = 7$ from $(7, 9)$.
Ans. $\sqrt{43}$.

5. Find that member of the family of circles through the intersection of $x^2 + y^2 - x - 2 = 0$ and $x^2 + y^2 + 5y - 1 = 0$ which passes through $(1, 1)$.
Ans. $7x^2 + 7y^2 - 6x + 5y - 13 = 0$.

6. Find the equation of the radical axis of the two circles $x^2 + y^2 - 16 = 0$ and $(x - 1)^2 + y^2 - 1 = 0$.
Ans. $x = 8$.

7. Show that the radical center of the three circles $x^2 + y^2 + x - y - 2 = 0$, $x^2 + y^2 - 7x + 5y - 8 = 0$, and $x^2 + y^2 = 3$ is the point $(0, 1)$.

8. Show that the two circles $x^2 + y^2 - 4x - 1 = 0$ and $x^2 + y^2 + x - y - 1 = 0$ are orthogonal.

9. Show how to construct the radical axis of two non-intersecting circles. (*Hint:* Draw another circle cutting each of the given circles and find the radical center.)

CHAPTER VI

THE PARABOLA

40. Definitions. *A parabola is the locus of a point P which moves so that the ratio of its distance from a fixed point and from a fixed line is unity.* The fixed point, F, is called the *focus* and the fixed line, D, the *directrix*. By definition, the distance from any point P on the parabola to the focus is equal to its distance to the directrix (the sign of the latter being disregarded). The ratio PF/PD is called the *eccentricity* e, and $e = 1$. The line FD through the focus perpendicular to the directrix is called the *axis* of the parabola. The midpoint V of the *segment FD*, obviously a point on the locus, is called the *vertex* of the parabola. The focal chord perpendicular to the axis is called the *latus rectum.*

41. General Equation of a Parabola. We choose any point $F(x_1, y_1)$ as focus and any line $\lambda x + \mu y - p = 0$ as directrix.

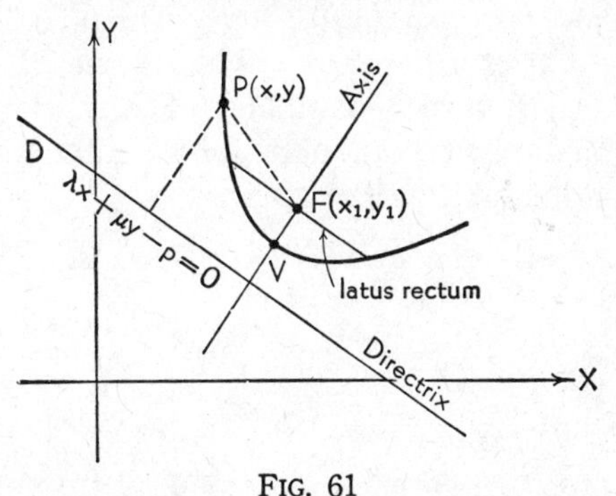

FIG. 61

The normal form of the line is used since distance is involved. With reference to these defining elements (Fig. 61) the equation of the parabola becomes

$$\sqrt{(x - x_1)^2 + (y - y_1)^2} = |\lambda x + \mu y - p|$$

which simplifies to

$$(1)\quad (\lambda^2 - 1)x^2 + 2\lambda\mu xy + (\mu^2 - 1)y^2 + 2(x_1 - \lambda p)x + 2(y_1 - \mu p)y + (p^2 - x_1^2 - y_1^2) = 0.$$

This is of the form $Ax^2 + Bxy + Cy^2 + Dx + Ey + F = 0$, an equation of the second degree. Moreover, it can be checked that $B^2 - 4AC \equiv 4\lambda^2\mu^2 - 4(\lambda^2 - 1)(\mu^2 - 1) \equiv 0$.

Therefore *a necessary condition that* $Ax^2 + Bxy + Cy^2 + Dx + Ey + F = 0$ *represent a parabola is that* $B^2 - 4AC = 0$. (See Chapter XI for a fuller discussion of this condition.)

Equation (1) reveals that if the directrix line is parallel to one of the coordinate axes then $B = 0$ since either λ or μ will be zero. In the equation of a parabola so placed there will therefore be no xy-term. This is a noteworthy remark.

42. Standard Forms of the Equation of a Parabola. The definition of a parabola makes the "shape" of the locus depend only upon the distance from focus to directrix and not essentially upon the coordinate system. Equation (1) is complicated because of the choice of a general point and a general line. By an appropriate choice of axes this equation can be simplified; but it will then represent only parabolas in special positions. For example if axes are chosen so that the focus has coordinates $(p, 0)$ and the directrix the equation $x = -p$, the locus definition yields $x + p = \sqrt{(x - p)^2 + y^2}$, which reduces to the very simple equation $y^2 = 4px$. This is one of the *standard forms* of the equation of a parabola. For quick reference we list the four fundamental standard forms, tabulating pertinent information and exhibiting the graphs.

I. *Axis Parallel to the X-Axis*

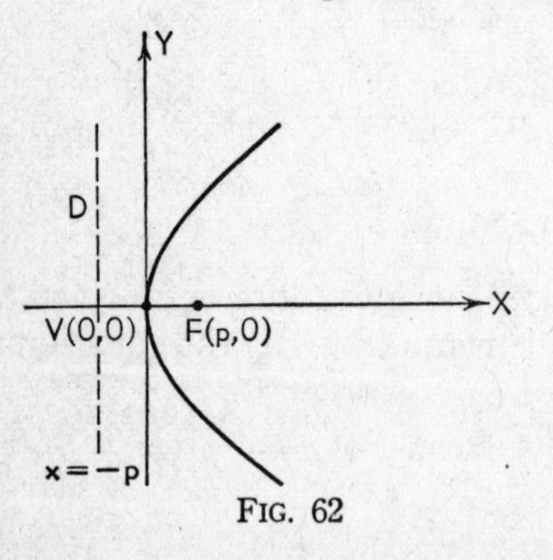

FIG. 62

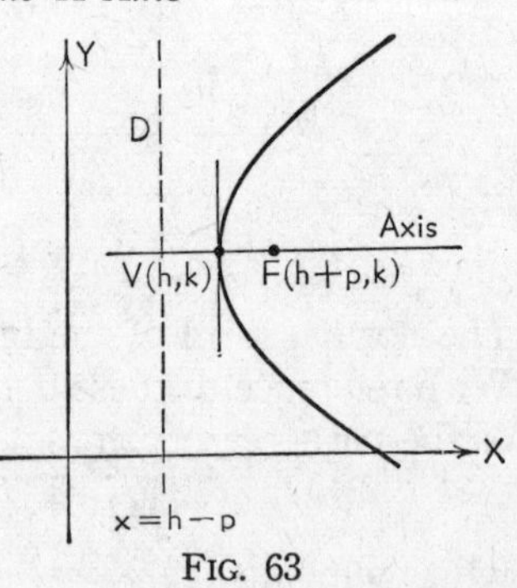

FIG. 63

Equation:	$y^2 = 4\,px$	$(y - k)^2 = 4\,p(x - h)$
Coordinates of vertex:	$V(0, 0)$	$V(h, k)$
Coordinates of focus:	$F(p, 0)$	$F(h + p, k)$
Equation of directrix:	$x = -p$	$x = h - p$
Length of latus rectum:	$\lvert 4\,p \rvert$	$\lvert 4\,p \rvert$

If p is positive the parabola extends to the right as in the figures; if p is negative it extends to the left.

II. *Axis Parallel to the Y-Axis*

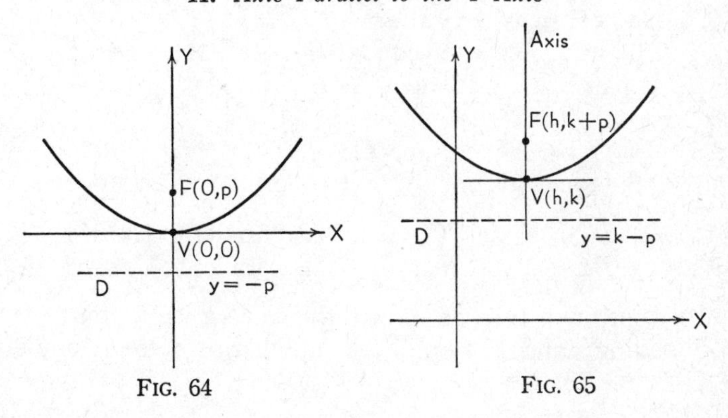

FIG. 64 FIG. 65

Equation:	$x^2 = 4\,py$	$(x - h)^2 = 4\,p(y - k)$
Coordinates of vertex:	$V(0, 0)$	$V(h, k)$
Coordinates of focus:	$F(0, p)$	$F(h, k + p)$
Equation of directrix:	$y = -p$	$y = k - p$
Length of latus rectum:	$\lvert 4\,p \rvert$	$\lvert 4\,p \rvert$

If p is positive the parabola extends upward as in the figures; if p is negative it extends downward. The vertex is respectively the *minimum* or *maximum* point on the curve.

A parabola is symmetric with respect to its axis. The numerical value of p is the distance between the vertex and focus. It is also the distance from vertex to directrix.

Illustration 1. Write the equation of the parabola for which $y = 1$ is the directrix and $F(3, -2)$ is the focus.

Solution. Since $\lvert 2\,p \rvert = 3$, $\lvert p \rvert = \frac{3}{2}$. The parabola turns down and $p = -\frac{3}{2}$. The vertex is at $(3, -\frac{1}{2})$ and the equation is

$$(x - 3)^2 = -6(y + \tfrac{1}{2}).$$

Illustration 2. (a) Write the equation of the parabola with vertex at $V(2, 3)$ and focus at $F(0, 3)$. (b) Find the latus rectum.

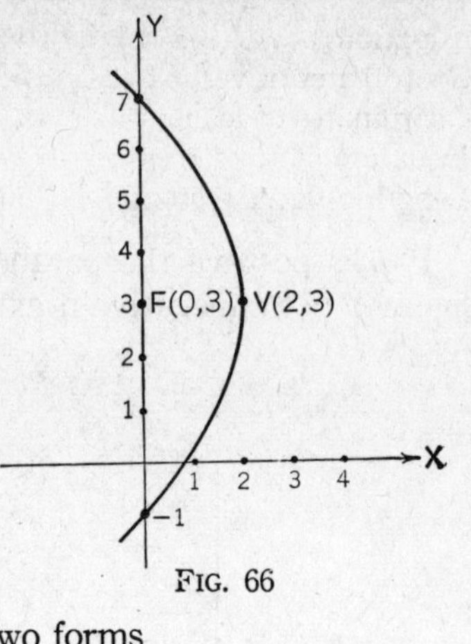

FIG. 66

Solution. (a) Since $p = -2$ the equation is

$$(y - 3)^2 = -8(x - 2).$$

(b) The latus rectum is $|4p| = 8$.

43. Reduction to Standard Form. The most general equation of a parabola with no xy-term present (and hence one whose axis is parallel to one of the coordinate axes) is one of the two forms

(1) $Ax^2 + Dx + Ey + F = 0,$ Axis parallel to the Y-axis;
(2) $Cy^2 + Dx + Ey + F = 0,$ Axis parallel to the X-axis.

In either case it is easy to reduce this general equation to the corresponding standard form by the familiar process of completing the square.

Illustration 1. Reduce $2x^2 - 3x + 8y + 1 = 0$ to standard form and sketch.

Solution. First we divide the given equation by 2 in order to reduce the coefficient of x^2 to unity.

$$x^2 - \frac{3x}{2} + 4y + \frac{1}{2} = 0.$$

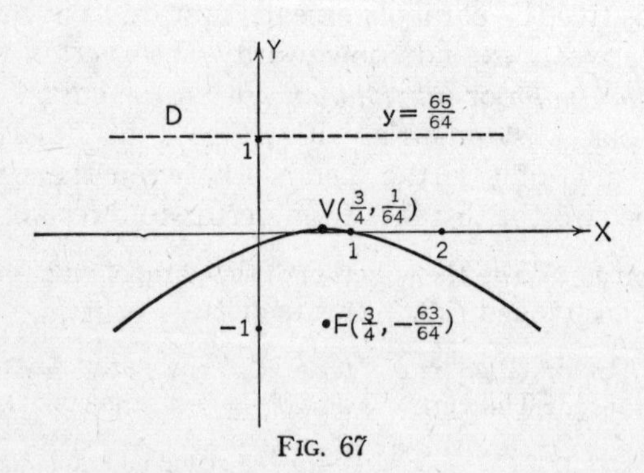

FIG. 67

Completing the square we have

$$x^2 - \frac{3x}{2} + \frac{9}{16} + 4y + \frac{1}{2} - \frac{9}{16} = 0,$$
$$(x - \tfrac{3}{4})^2 = -4y + \tfrac{1}{16},$$
$$(x - \tfrac{3}{4})^2 = -4(y - \tfrac{1}{64}).$$

To make an accurate sketch we should indicate the vertex, focus, and directrix. From the equation these are $V(\frac{3}{4}, \frac{1}{64})$, $F(\frac{3}{4}, -\frac{63}{64})$, and $y = \frac{65}{64}$.

Illustration 2. Write the equation of the axis of the parabola $y^2 - 4x + 4y + 7 = 0$.

Solution. Reduced to standard form the equation is

$$(y + 2)^2 = 4(x - \tfrac{3}{4}),$$

giving $V(\frac{3}{4}, -2)$, $p = 1$, $F(\frac{7}{4}, -2)$. Since the axis passes through V and F its equation is $y = -2$.

44. Equation of a Tangent. A line which is parallel to the axis of a parabola intersects the parabola in only one (finite) point; all other lines will cut the parabola in two points real and distinct, real and coincident, or complex. A line which meets a parabola in two coincident points is called a *tangent.* (A tangent to any curve at a point P is the *limiting position* of a secant line, cutting the curve in two points P and Q, as $Q \to P$.)

We summarize the formulae for tangents to parabolas.

PARABOLA	TANGENT	
	At (x_1, y_1)	With Slope m
$y^2 = 4px$	$yy_1 = 2p(x + x_1)$	$y = mx + p/m$
$x^2 = 4py$	$xx_1 = 2p(y + y_1)$	$y = mx - pm^2$
$(y-k)^2 = 4p(x-h)$	$(y-k)(y_1-k) = 2p(x + x_1 - 2h)$	$y - k = m(x-h) + p/m$
$(x-h)^2 = 4p(y-k)$	$(x-h)(x_1-h) = 2p(y + y_1 - 2k)$	$y - k = m(x-h) - pm^2$

Illustration 1. Write the equation of (a) the tangent and (b) the normal to the parabola $y^2 - 12x - 2y - 23 = 0$ at $(1, 7)$.

Solution. (a) In standard form the equation of the parabola is

$$(y - 1)^2 = 12(x + 2);$$

the equation of the tangent is

$$(y - 1)(7 - 1) = 6(x + 1 + 4),$$

or

$$x - y + 6 = 0.$$

(b) The equation of the normal will therefore be

$$y - 7 = -(x - 1),$$

or $\quad x + y - 8 = 0.$

Illustration 2. Write the equation of the line of slope 4 which is tangent to $x^2 = y$.

Solution. The tangent is

$$y = 4x - (\tfrac{1}{4})(4)^2,$$

or $\quad y = 4x - 4.$

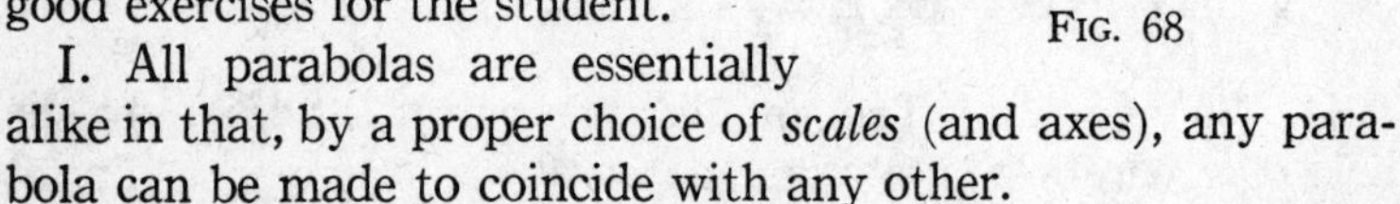

FIG. 68

45. Properties of a Parabola. The following properties are listed without proof. They will serve as good exercises for the student.

I. All parabolas are essentially alike in that, by a proper choice of *scales* (and axes), any parabola can be made to coincide with any other.

II. Let P be any point on a parabola and let Q be the projection of P on the axis of the parabola. Then

$$\overline{PQ}^2 = 4(VF)(VQ).$$

This fundamental property can be used to write down the standard forms of the equation. Indeed it can be used to define a parabola. (Fig. 69.)

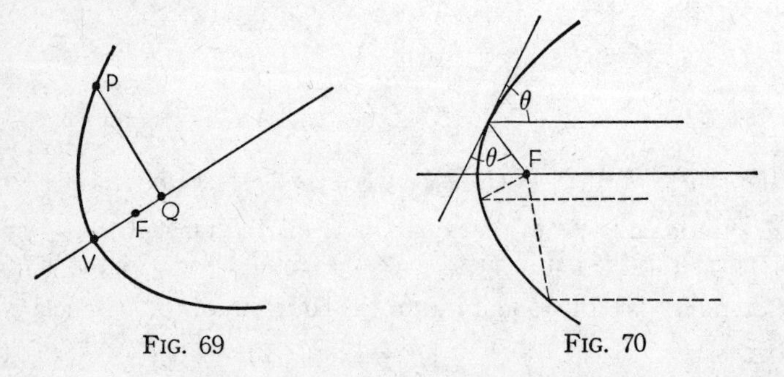

FIG. 69 FIG. 70

III. A tangent to a parabola makes equal angles with the axis of the parabola and the focal chord drawn to the point of tangency. This is the well-known reflective property of parabolic searchlights and is the source of the name "focus." (Fig. 70.)

IV. A tangent and a line perpendicular to it through the focus intersect on the tangent at the vertex. (Fig. 71.)

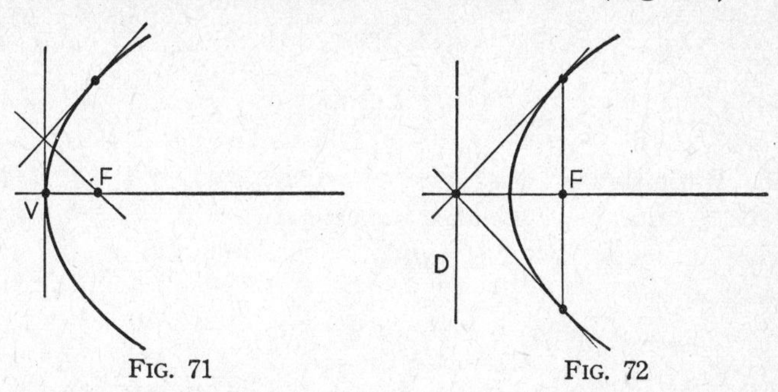

FIG. 71 FIG. 72

V. The tangents at the ends of the latus rectum, the directrix, and the axis are concurrent. (Fig. 72.)

EXERCISES

Reduce the following equations to standard form; write the coordinates of vertex and focus; write the equations of the directrix and axis.

1. $x^2 - 3y + 3 = 0$.

Ans. $x^2 = 3(y - 1)$; $V(0, 1)$, $F(0, \frac{7}{4})$; $y = \frac{1}{4}$, $x = 0$.

2. $y^2 - 8x - 8y + 8 = 0$.

Ans. $(y - 4)^2 = 8(x + 1)$; $V(-1, 4)$, $F(1, 4)$; $x = -3$, $y = 4$.

3. $4x^2 + 4x + 4y + 9 = 0$.

Ans. $(x + \frac{1}{2})^2 = -(y + 2)$; $V(-\frac{1}{2}, -2)$, $F(-\frac{1}{2}, -\frac{9}{4})$; $y = -\frac{7}{4}$, $x = -\frac{1}{2}$.

Find the equation of the tangent to

4. $x^2 - 4y + 7 = 0$ at $(2, \frac{11}{4})$. *Ans.* $4x - 4y + 3 = 0$.

5. $(y - 2)^2 = -3(x + 1)$ with slope $\frac{1}{2}$. *Ans.* $x - 2y + 2 = 0$.

6. Find the locus of a point which moves so that it is always equidistant from $3x - 4y + 7 = 0$ and $(1, -2)$.

Ans. $16x^2 + 24xy + 9y^2 - 92x + 156y + 76 = 0$.

7. A variable circle S is always tangent to $x = -1$ and passes through $(1, 0)$. Find the locus of the center of S. *Ans.* $y^2 = 4x$.

CHAPTER VII

THE ELLIPSE

46. Definitions. *An ellipse is the locus of a point which moves so that the ratio of its distance from a fixed point and from a fixed line is a constant less than unity.* The distances involved are numerical.

Fixed point F	Focus
Fixed line D	Directrix
Fixed constant e (<1)	Eccentricity
$\dfrac{PF}{PD} = e$	Definition of ellipse

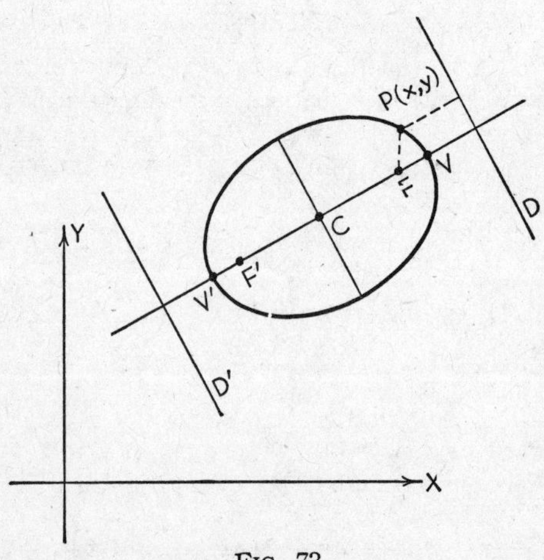

FIG. 73

Consider the line through the focus perpendicular to the directrix. From the definition $PF/PD = e/1$ there are obviously two points V and V' which divide the (undirected) segment FD, internally and externally respectively, in the ratio of $e/1$. Therefore V and V' are points (on the *same* side of D) on the ellipse; they are called the *vertices*. The segment VV'

is called the *major axis.* By symmetry there is another point F' and another line D' such that F' and D' would serve in the definition of this curve. Thus an *ellipse has two foci and two directrices* associated in pairs F, D and F', D'. The midpoint of FF', which is also the midpoint of VV', is called the *center* C of the ellipse. It is evident that the locus is contained between the vertices, that it is bounded in all directions, and that it is symmetric both with respect to the major axis and to a line perpendicular to it through C. It is therefore a closed curve (an oval).

The length of the focal chord perpendicular to the major axis is called the *latus rectum.* The length of the central chord perpendicular to the major axis is called the *minor axis.*

47. General Equation of an Ellipse. Take any point $F(x_1, y_1)$ and any line D, $\lambda x + \mu y - p = 0$. By definition the equation of the ellipse is

$$\sqrt{(x - x_1)^2 + (y - y_1)^2} = e|\lambda x + \mu y - p|$$

which reduces to

$$(1) \quad (e^2\lambda^2 - 1)x^2 + 2\, e^2\lambda\mu xy + (e^2\mu^2 - 1)y^2 + 2(x_1 - e^2\lambda p)x + 2(y_1 - e^2\mu p)y + (e^2p^2 - x_1^2 - y_1^2) = 0.$$

This is of the form $Ax^2 + Bxy + Cy^2 + Dx + Ey + F = 0$, an equation of the second degree. Moreover it can be checked that $B^2 - 4\,AC \equiv 4(e^2 - 1) < 0$ (when $e < 1$).

Therefore *a necessary condition that* $Ax^2 + Bxy + Cy^2 + Dx + Ey + F = 0$ *represent an ellipse is that* $B^2 - 4\,AC < 0$. (See Chapter XI on the general equation of second degree for a fuller discussion of this condition.)

Equation (1) reveals that if the defining directrix line is parallel to one of the coordinate axes then $B = 0$, since either λ or μ will be zero. In the equation of an ellipse so placed there will therefore be no xy-term. This is noteworthy.

48. Standard Forms of the Equation of an Ellipse. By an appropriate choice of axes the general equation can be reduced to one of the following *standard forms.*

I. *Major Axis Parallel to the X-Axis*

a = semimajor axis

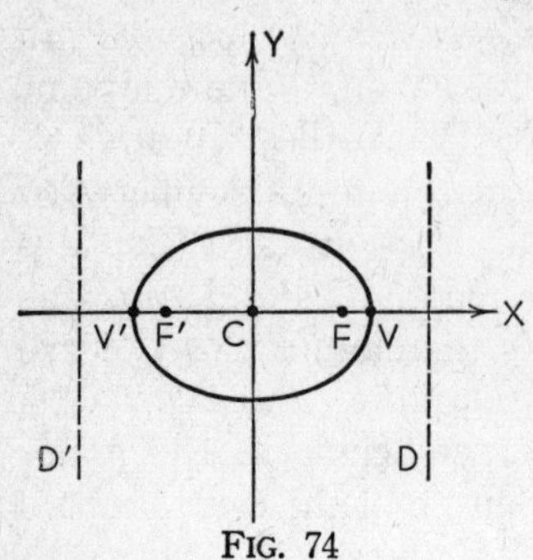

FIG. 74

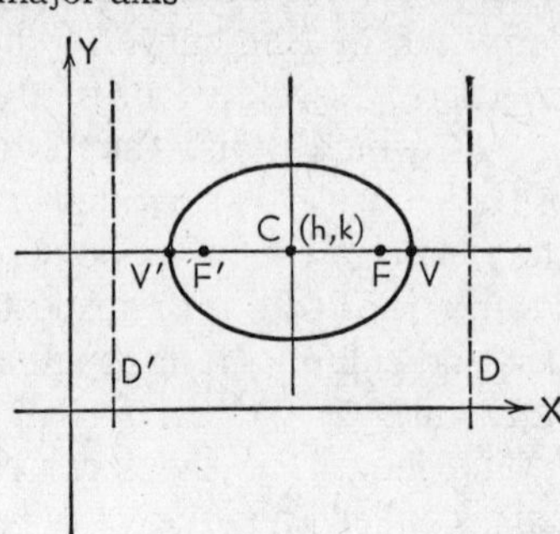

FIG. 75

Equation:	$\frac{x^2}{a^2} + \frac{y^2}{b^2} = 1$	$\frac{(x - h)^2}{a^2} + \frac{(y - k)^2}{b^2} = 1$
Coordinates of vertices:	$V(a, 0)$, $V'(-a, 0)$	$V(h + a, k)$, $V'(h - a, k)$
Coordinates of foci:	$F(ae, 0)$, $F'(-ae, 0)$	$F(h + ae, k)$, $F'(h - ae, k)$
Coordinates of center:	$C(0, 0)$	$C(h, k)$
Equations of directrices.	$x = \pm \frac{a}{e}$	$x = h \pm \frac{a}{e}$

Semimajor axis: a

Semiminor axis: b

Eccentricity: $e = \frac{\sqrt{a^2 - b^2}}{a} < 1$

Length of latus rectum: $\frac{2\,b^2}{a}$

II. *Major Axis Parallel to the Y-Axis*

a = semimajor axis

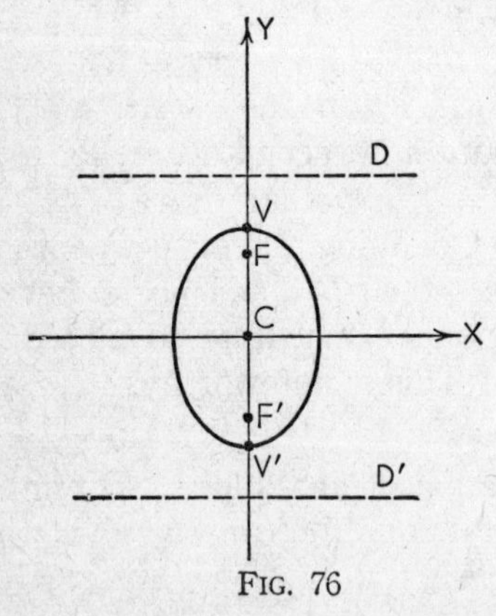

FIG. 76

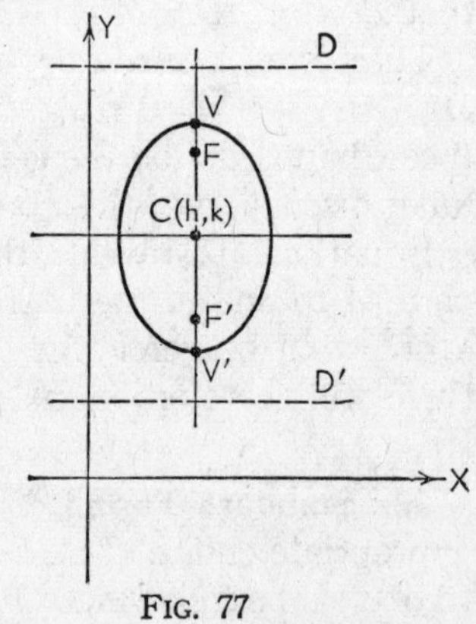

FIG. 77

Equation:	$\frac{x^2}{b^2} + \frac{y^2}{a^2} = 1$	$\frac{(x-h)^2}{b^2} + \frac{(y-k)^2}{a^2} = 1$
Coordinates of vertices:	$V(0, a)$, $V'(0, -a)$	$V(h, k+a)$, $V'(h, k-a)$
Coordinates of foci:	$F(0, ae)$, $F'(0, -ae)$	$F(h, k+ae)$, $F'(h, k-ae)$
Coordinates of center:	$C(0, 0)$	$C(h, k)$
Equations of directrices:	$y = \pm \frac{a}{e}$	$y = k \pm \frac{a}{e}$

Semimajor axis: a

Semiminor axis: b

Eccentricity: $e = \frac{\sqrt{a^2 - b^2}}{a} < 1$

Length of latus rectum: $\frac{2\,b^2}{a}$

Note: There is a minimum number of changes in the formulae, and these are natural and easy to remember, if we continue to use a as the semimajor axis. Thus the *larger* of the two numbers under x^2 and y^2 in a given equation is to be thought of as a^2.

Illustration 1. (a) Write the equation of the ellipse with center at $(2, -1)$, with major axis $= 10$ and parallel to the X-axis, and with minor axis $= 8$. (b) Find the eccentricity, the foci, and the vertices. (c) Write the equations of the directrices.

Solution. (a) The equation is

$$\frac{(x-2)^2}{25} + \frac{(y+1)^2}{16} = 1.$$

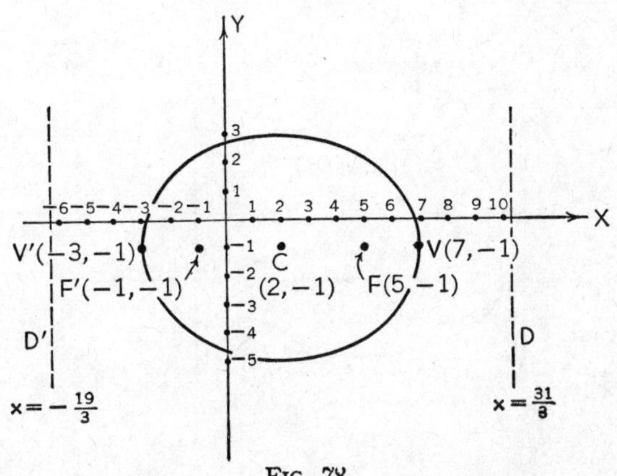

FIG. 78

(b) $e = \dfrac{\sqrt{25 - 16}}{5} = \dfrac{3}{5}$, $ae = 3$, $F(2 + 3, -1)$, $F'(2 - 3, -1)$ or $F(5, -1)$, $F'(-1, -1)$, $V(7, -1)$, $V'(-3, -1)$.

(c) $$x = 2 \pm \frac{5}{\frac{3}{5}} \quad \text{or} \quad x = \tfrac{31}{3},\ x = -\tfrac{19}{3}.$$

Illustration 2. Sketch the ellipse $2x^2 + y^2 = 4$.

Solution. In standard form the equation is

$$\frac{x^2}{2} + \frac{y^2}{4} = 1.$$

We compute: $a = 2$, $b = \sqrt{2}$, $e = \frac{1}{2}\sqrt{2}$, $C(0, 0)$, $V(0, 2)$, $V'(0, -2)$, $F(0, \sqrt{2})$, $F'(0, -\sqrt{2})$. The major axis is parallel to (coincides with) the Y-axis.

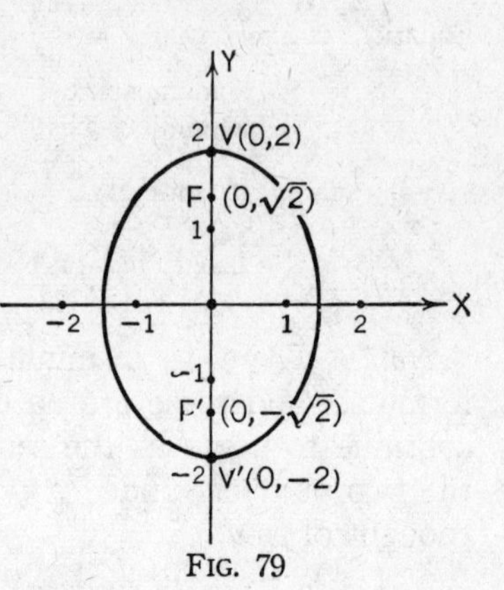

FIG. 79

49. Reduction to Standard Form. The most general equation of an ellipse with no xy-term present (and hence one whose axes are parallel to the coordinate axes) is of the form

$$(1) \qquad Ax^2 + Cy^2 + Dx + Ey + F = 0, \quad AC > 0.$$

The condition $B^2 - 4AC < 0$ reduces to $AC > 0$ in this case, since $B = 0$; it implies that A and C are of like sign. This equation can be reduced to one of the standard forms by completing the square.

Illustration 1. Reduce $x^2 + 4y^2 + 4x = 0$ to standard form and sketch.

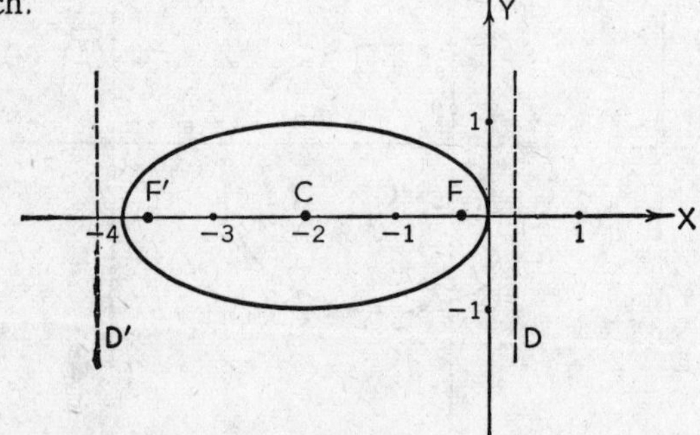

FIG. 80

Solution.

$$x^2 + 4x + 4 + 4y^2 = 4,$$
$$(x + 2)^2 + 4y^2 = 4,$$
$$\frac{(x + 2)^2}{4} + \frac{y^2}{1} = 1. \quad \text{(Standard form)}$$

Further, $a = 2$, $b = 1$, $e = \frac{\sqrt{3}}{2}$, $C(-2, 0)$, $V(0, 0)$, $V'(-4, 0)$, $F(-2 + \sqrt{3}, 0)$, $F'(-2 - \sqrt{3}, 0)$. The directrices have the equations $x = -2 \pm \frac{4}{3}\sqrt{3}$ and the latus rectum is 1. The major axis is parallel to (coincides with) the X-axis.

Illustration 2. Reduce $5x^2 - 10x + 9y^2 - 54y + 41 = 0$ to standard form.

Solution. We reduce the coefficients of x^2 and y^2 within parenthesis to unity by factoring out of the x-terms and the y-terms 5 and 9 respectively, obtaining

$$5(x^2 - 2x \quad\quad) + 9(y^2 - 6y \quad\quad) = -41.$$

Completing the squares we get

$$5(x^2 - 2x + 1) + 9(y^2 - 6y + 9) = -41 + 5 + 81.$$

Note that we have actually added 5 and 81 to each side of the equation. Finally we obtain as the standard form of the equation

$$\frac{(x - 1)^2}{9} + \frac{(y - 3)^2}{5} = 1.$$

Further $a = 3$, $b = \sqrt{5}$, $e = \frac{2}{3}$, $V(4, 3)$, $V'(-2, 3)$; with this information it would be easy to sketch the graph. This is left as an exercise for the student.

50. Equation of a Tangent. A line which intersects an ellipse in two coincident points is a tangent. As in the case of the circle, but unlike that of the parabola, there will be two tangents to an ellipse with a given slope. We summarize the formulae for tangents to an ellipse.

Ellipse

$$\frac{x^2}{a^2} + \frac{y^2}{b^2} = 1$$

$$\frac{(x - h)^2}{a^2} + \frac{(y - k)^2}{b^2} = 1$$

TANGENT

At (x_1, y_1)	With Slope m
$\frac{xx_1}{a^2} + \frac{yy_1}{b^2} = 1$	$y = mx \pm \sqrt{a^2m^2 + b^2}$
$\frac{(x-h)(x_1-h)}{a^2} + \frac{(y-k)(y_1-k)}{b^2} = 1$	$y-k=m(x-h) \pm \sqrt{a^2m^2+b^2}$

Note: Remember that the larger of the two numbers under x^2 and y^2 is to be called a^2. If a^2 is under y^2 or $(y - k)^2$, the a and the b in the formulae for the tangent will have to be interchanged.

Illustration 1. Write the equation of (a) the tangent and (b) the normal to the ellipse $16x^2 + y^2 - 16 = 0$ at $(\frac{1}{2}, -2\sqrt{3})$.

Solution. (a) The equation of the tangent is

$$\frac{x(\frac{1}{2})}{1} + \frac{y(-2\sqrt{3})}{16} = 1,$$

or

$$4x - \sqrt{3}\,y - 8 = 0.$$

(b) The equation of the normal is

$$y + 2\sqrt{3} = -\frac{\sqrt{3}}{4}(x - \tfrac{1}{2}).$$

Illustration 2. Write the equations of the lines of slope 2 which are tangent to

$$13x^2+3y^2-26x+24y+22=0.$$

Solution. The equation of the ellipse in standard form is

$$\frac{(x-1)^2}{3} + \frac{(y+4)^2}{13} = 1.$$

The equations of the tangents are

$$y + 4 = 2(x - 1) \pm 5,$$

which reduce to

$$2x - y - 1 = 0$$

and

$$2x - y - 11 = 0.$$

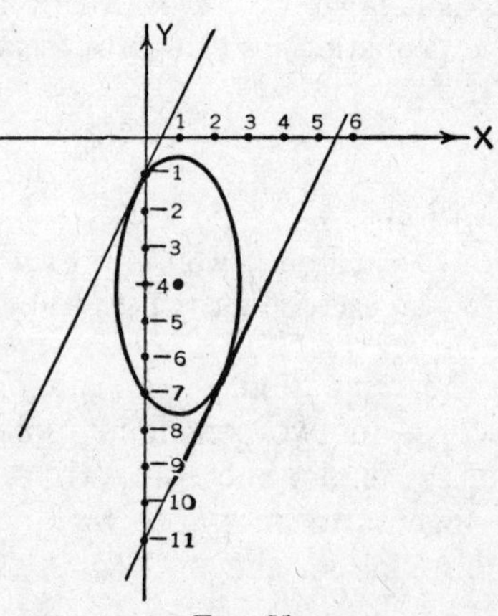

FIG. 81

51. Properties of an Ellipse. The following properties are listed without proof. They will serve as good exercises for the student.

I. All ellipses of like eccentricity are essentially alike and by a proper choice of *scales* (and axes) can be made to coincide. But ellipses of unlike eccentricity are unlike in "shape."

II. Let P be any point on an ellipse and let Q and R be the projections of P onto the major and minor axes ($2a$ and $2b$) respectively. Then

$$\frac{\overline{PR}^2}{a^2} + \frac{\overline{PQ}^2}{b^2} = 1.$$

This fundamental property can be used to write down the standard forms of the equation. Indeed, it can be used to define an ellipse. (Fig. 82.)

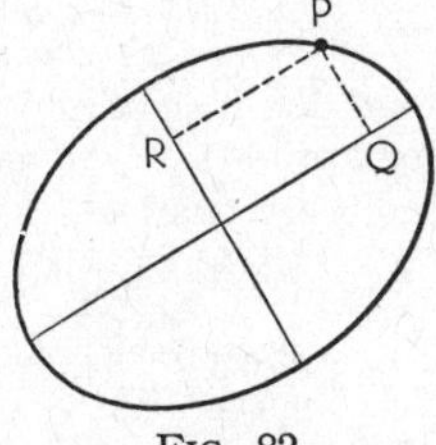

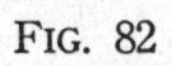
FIG. 82

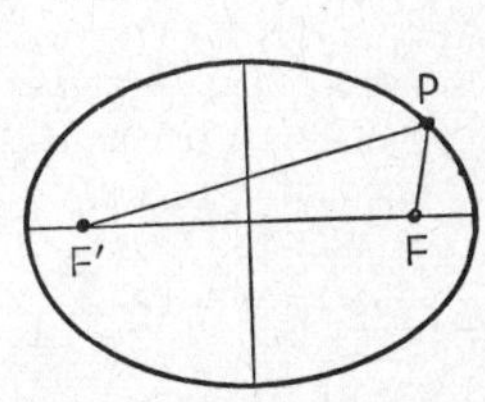

FIG. 83

III. The sum of the focal radii for any point P on an ellipse is constant and equal to the major axis:

$$PF' + PF = 2a.$$

This also is a fundamental property and can be used as a definition of the ellipse. Thus *an ellipse is the locus of a point which moves so that the sum of its distances from two fixed points is a constant.* (Fig. 83.)

IV. A tangent to an ellipse makes equal angles with the focal radii drawn to the point of tangency. This expresses the well-known optical property of an ellipse and is the source of the name "foci." (Fig. 84.)

V. The perpendicular from the focus F upon the tangent at P will meet the line joining the center and P on the directrix corresponding to F. (Fig. 85.)

VI. The perpendiculars from the foci to any tangent meet the tangent in points which lie on the *auxiliary circle* $x^2 + y^2 = a^2$. (Fig. 85.)

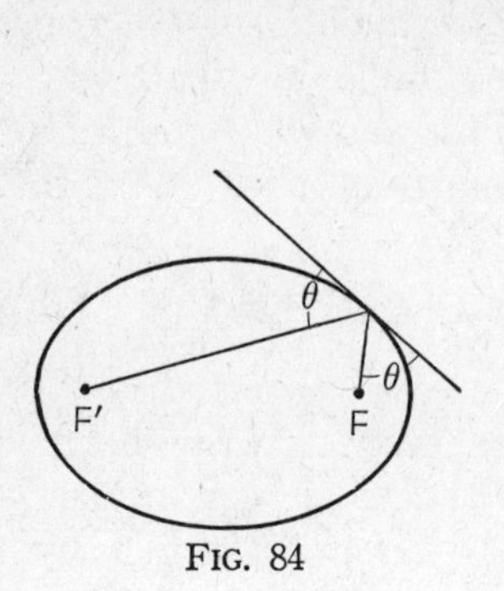

FIG. 84

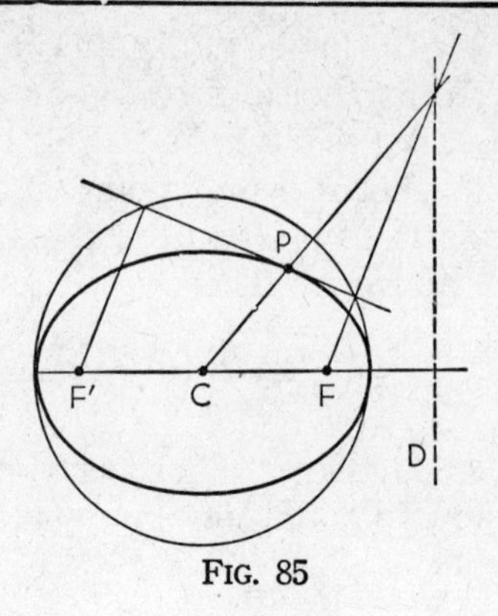

FIG. 85

EXERCISES

Reduce the following equations to standard form; compute the semimajor and semiminor axes and the eccentricity; write the coordinates of the center, vertices, and foci; write the equations of the directrices.

1. $7x^2 + 4y^2 - 14x + 40y + 79 = 0.$

Ans. $\dfrac{(x-1)^2}{4} + \dfrac{(y+5)^2}{7} = 1;\quad a = \sqrt{7},\quad b = 2,\quad e = \dfrac{\sqrt{3}}{\sqrt{7}};\quad C(1, -5),$
$V(1, -5 + \sqrt{7}),\ V'(1, -5 - \sqrt{7}),\ F(1, -5 + \sqrt{3}),$
$F'(1, -5 - \sqrt{3});\ y = -5 \pm \frac{7}{3}\sqrt{3}.$

2. $9x^2 + 16y^2 + 36x - 32y - 92 = 0.$

Ans. $\dfrac{(x+2)^2}{16} + \dfrac{(y-1)^2}{9} = 1;\quad a = 4,\quad b = 3,\quad e = \frac{1}{4}\sqrt{7};\quad C(-2, 1),$
$V(2, 1),\ V'(-6, 1),\ F(-2 + \sqrt{7}, 1),\ F'(-2 - \sqrt{7}, 1);$
$x = -2 \pm \frac{16}{7}\sqrt{7}.$

Find the equation(s) of the tangent(s) to

3. $x^2 + 5y^2 - 3x + 2y = 0$ at $(3, 0)$. *Ans.* $3x + 2y - 9 = 0.$

4. $16x^2 + 9y^2 - 32x + 54y - 47 = 0$ with slope -1.

Ans. $x + y + 7 = 0$ and $x + y - 3 = 0.$

5. Find the locus of a point which moves so that the sum of its distances from $(-3, 0)$ and $(3, 0)$ always equals 8.

Ans. $\dfrac{x^2}{16} + \dfrac{y^2}{7} = 1.$

CHAPTER VIII

THE HYPERBOLA

52. Definitions. *A hyperbola is the locus of a point P which moves so that the ratio of its distance from a fixed point and from a fixed line is a constant greater than unity.* The distances involved are numerical.

Fixed point F	Focus
Fixed line D	Directrix
Fixed constant e (>1)	Eccentricity
$\dfrac{PF}{PD} = e$	Definition of hyperbola

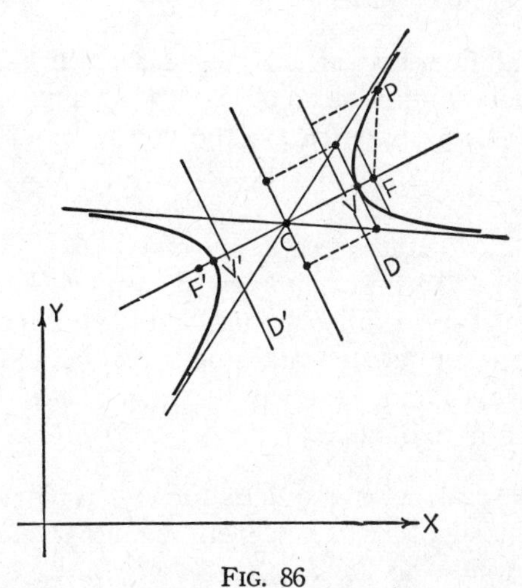

FIG. 86

Consider the line through the focus perpendicular to the directrix. From the definition $PF/PD = e/1$ there are two points V and V' which divide the (undirected) segment FD, internally and externally respectively, in the ratio of $e/1$. Therefore V and V' are points (on *opposite* sides of D) on the hyperbola; they are called the *vertices*. The segment VV' is called the *transverse axis*. By symmetry there is another point F' and another line D' such that F' and D' would serve in defining this curve. Thus a *hyperbola has two foci and two directrices* associated in pairs F, D and F', D'. The midpoint of FF', which is also the midpoint of VV', is called the *center* C of the hyperbola. There is no part of the locus between the vertices, each being on a separate branch extending to infinity. There is symmetry with respect to C. There are two tangents through C whose points of contact are at an infinite distance from C. These are called the *asymptotes* of the hyperbola. (See the discussion, p. 34.) The length of the focal chord perpendicular to the transverse axis (extended) is called the *latus rectum*. A line through C perpendicular to the transverse axis does not intersect the hyperbola in real points. But the portion of it, bisected at C, which is equal in length to the parallel segment through V contained between the asymptotes is called the *conjugate axis*.

53. General Equation of a Hyperbola. Since the only difference in the definition of an ellipse and a hyperbola pertains to the value of the eccentricity, the general equation of each is the same. [See (1), § 47, p. 81.] But for the hyperbola $e > 1$ and $B^2 - 4AC \equiv 4(e^2 - 1) > 0$.

Therefore *a necessary condition that* $Ax^2 + Bxy + Cy^2 + Dx + Ey + F = 0$ *represent a hyperbola is that* $B^2 - 4AC > 0$. (See Chapter XI for a fuller discussion of this condition.)

Again, if the defining directrix line is parallel to a coordinate axis, B will be zero and there will be no xy-term. The student should make a note of this.

54. Standard Forms of the Equation of a Hyperbola. By an appropriate choice of axes the general equation can be reduced to one of the following *standard forms*.

I. *Transverse Axis Parallel to the X-Axis*

a = semitransverse axis

Equation:	$\frac{x^2}{a^2} - \frac{y^2}{b^2} = 1$	$\frac{(x-h)^2}{a^2} - \frac{(y-k)^2}{b^2} = 1$
Coordinates of vertices:	$V(a, 0)$, $V'(-a, 0)$	$V(h+a, k)$, $V'(h-a, k)$

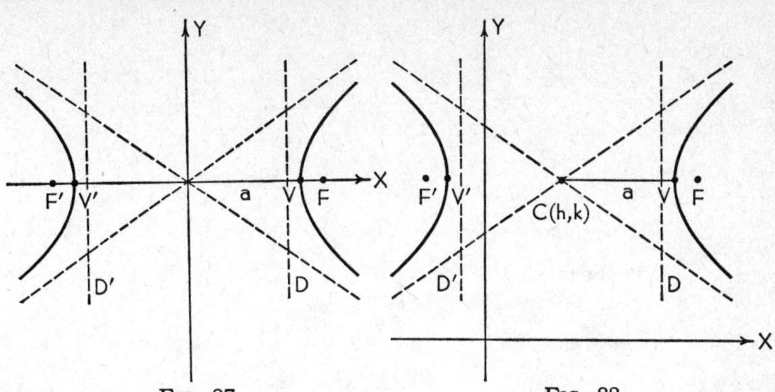

FIG. 87 FIG. 88

Coordinates of foci:	$F(ae, 0), F'(-ae, 0)$	$F(h + ae, k), F'(h - ae, k)$
Coordinates of center:	$C(0, 0)$	$C(h, k)$
Equations of directrices:	$x = \pm \frac{a}{e}$	$x = h \pm \frac{a}{e}$
Equations of asymptotes:	$\frac{x^2}{a^2} - \frac{y^2}{b^2} = 0$	$\frac{(x - h)^2}{a^2} - \frac{(y - k)^2}{b^2} = 0$

Semitransverse axis: a
Semiconjugate axis: b
Eccentricity: $e = \frac{\sqrt{a^2 + b^2}}{a} > 1$
Length of latus rectum: $\frac{2\,b^2}{a}$

II. *Transverse Axis Parallel to the Y-Axis*

a = semitransverse axis

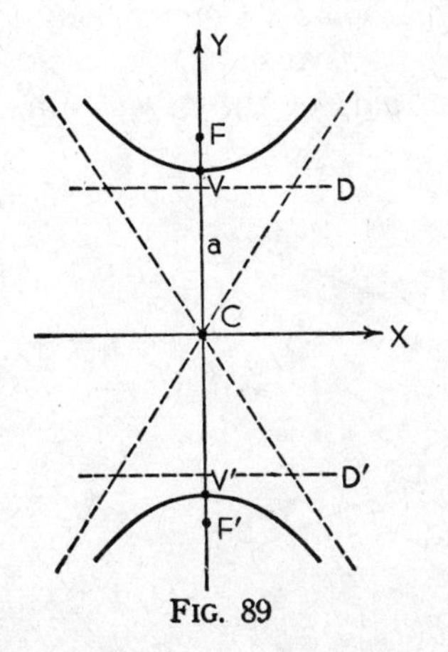

FIG. 89

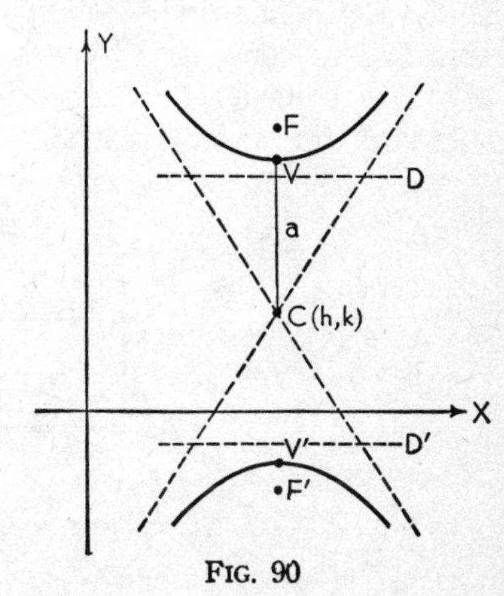

FIG. 90

Equation:	$-\frac{x^2}{b^2}+\frac{y^2}{a^2}=1$	$-\frac{(x-h)^2}{b^2}+\frac{(y-k)^2}{a^2}=1$
Coordinates of vertices:	$V(0, a)$, $V'(0, -a)$	$V(h, k+a)$, $V'(h, k-a)$
Coordinates of foci:	$F(0, ae)$, $F'(0, -ae)$	$F(h, k+ae)$, $F'(h, k-ae)$
Coordinates of center:	$C(0, 0)$	$C(h, k)$
Equations of directrices:	$y=\pm\frac{a}{e}$	$y=k\pm\frac{a}{e}$
Equations of asymptotes:	$-\frac{x^2}{b^2}+\frac{y^2}{a^2}=0$	$-\frac{(x-h)^2}{b^2}+\frac{(y-k)^2}{a^2}=0$

Semitransverse axis: a
Semiconjugate axis: b
Eccentricity: $e=\frac{\sqrt{a^2+b^2}}{a}>1$
Length of latus rectum: $\frac{2\,b^2}{a}$

Note: There is a minimum number of changes in the formulae, and these are natural and easy to remember, if we continue to use a as the semitransverse axis. Thus the denominator of the *positive term* in the standard form of the equation is to be thought of as a^2.

Note also that the equations of the asymptotes are obtained directly from the equation of the hyperbola by simply changing the right-hand member from unity to zero. The left-hand member of the equations of the asymptotes will factor into two linear factors. The presence of the asymptotes aids in the plotting of the hyperbola.

55. Conjugate and Rectangular Hyperbolas. *Conjugate hyperbolas* are concentric hyperbolas the transverse axis of each of which coincides with the conjugate axis of the other. In standard form their equations are

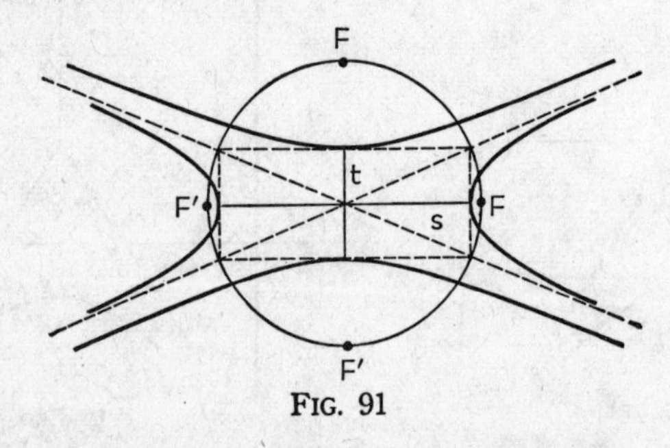

FIG. 91

	Center at (0, 0)	Center at (h, k)
Conjugate hyperbolas	$\frac{x^2}{s^2} - \frac{y^2}{t^2} = 1$	$\frac{(x-h)^2}{s^2} - \frac{(y-k)^2}{t^2} = 1$
	$-\frac{x^2}{s^2} + \frac{y^2}{t^2} = 1$	$-\frac{(x-h)^2}{s^2} + \frac{(y-k)^2}{t^2} = 1$

It is evident that if s is the semitransverse axis of one it is the semiconjugate axis of the other and vice versa. Conjugate hyperbolas have the same asymptotes and the foci lie on a circle with center at the center of the hyperbolas.

A *rectangular* (or *equilateral*) *hyperbola* is one in which the transverse and conjugate axes are equal, in which case the asymptotes are at right angles.

Illustration 1. (a) Write the equation of the hyperbola with center at $(-2, 1)$, with transverse axis $= 6$ and parallel to the X-axis, and with conjugate axis $= 8$. (b) Find the eccentricity, the foci, and the vertices. (c) Write the equations of the directrices and of the asymptotes.

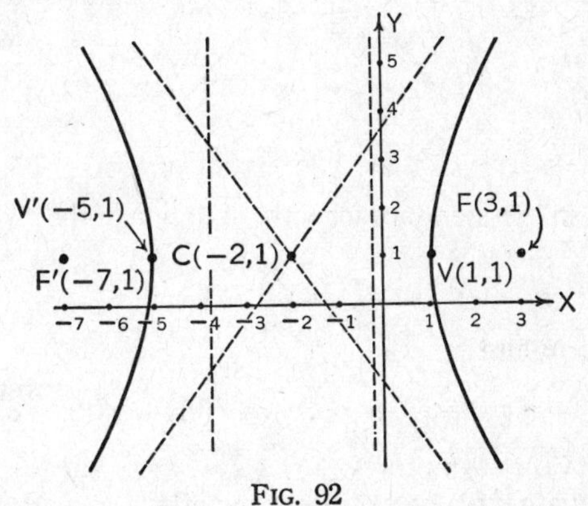

FIG. 92

Solution. (a) The equation is

$$\frac{(x+2)^2}{9} - \frac{(y-1)^2}{16} = 1.$$

(b) $e = \frac{\sqrt{9+16}}{3} = \frac{5}{3}$, $ae = 5$, $F(3, 1)$. $F'(-7, 1)$, $V(1, 1)$, $V'(-5, 1)$.

(c) Directrices: $x = -2 \pm \frac{9}{5}$, or $x = -\frac{1}{5}$, $x = -\frac{19}{5}$;

$$\text{Asymptotes: } \frac{(x+2)^2}{9} - \frac{(y-1)^2}{16} = 0, \text{ or } y-1 = \pm \frac{4}{3}(x+2).$$

Illustration 2. (a) Sketch the hyperbola $4x^2 - y^2 + 36 = 0$. (b) Write down the equation of the conjugate hyperbola and sketch.

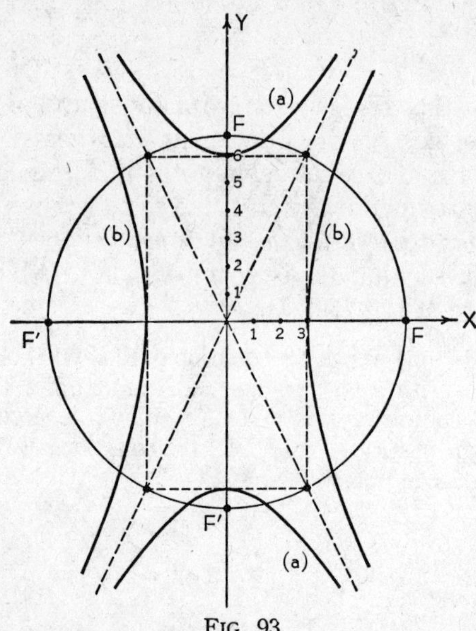

FIG. 93

Solution. (a) In standard form the equation of the given hyperbola is

$$-\frac{x^2}{9} + \frac{y^2}{36} = 1.$$

For it we compute

$a = 6, b = 3, e = \dfrac{\sqrt{5}}{2}, ae = 3\sqrt{5}, V(0, 6), V'(0, -6), F(0, 3\sqrt{5})$. $F'(0, -3\sqrt{5})$; asymptotes: $y = \pm 2x$.

(b) The equation of the conjugate hyperbola is

$$\frac{x^2}{9} - \frac{y^2}{36} = 1.$$

For it we compute

$a = 3, b = 6, e = \sqrt{5}, ae = 3\sqrt{5}, V(3, 0), V'(-3, 0), F(3\sqrt{5}, 0)$, $F'(-3\sqrt{5}, 0)$; asymptotes: $y = \pm 2x$.

56. Reduction to Standard Form. The most general equation of a hyperbola with no xy-term present (axes parallel to the coordinate axes) is of the form

$$(1) \qquad Ax^2 + Cy^2 + Dx + Ey + F = 0, \quad AC < 0.$$

The condition $B^2 - 4\,AC > 0$ reduces to $AC < 0$ in this case since $B = 0$; it implies that A and C are of opposite sign. This equation can be reduced to one of the standard forms by completing the square.

Illustration 1. Reduce $x^2 - y^2 - 2\,x - y + 1 = 0$ to standard form and sketch.

Solution.

$$x^2 - 2\,x + 1 - (y^2 + y + \tfrac{1}{4}) = -1 + 1 - \tfrac{1}{4},$$
$$(x-1)^2 - (y + \tfrac{1}{2})^2 = -\tfrac{1}{4},$$
$$-\frac{(x-1)^2}{\frac{1}{4}} + \frac{(y+\frac{1}{2})^2}{\frac{1}{4}} = 1.$$

This is a rectangular hyperbola with $a = b = \frac{1}{2}$, $e = \sqrt{2}$, $V(1, 0)$, $V'(1, -1)$, $F(1, -\frac{1}{2} + \frac{1}{2}\sqrt{2})$, $F'(1, -\frac{1}{2} - \frac{1}{2}\sqrt{2})$; asymptotes: $y + \frac{1}{2} = \pm(x - 1)$.

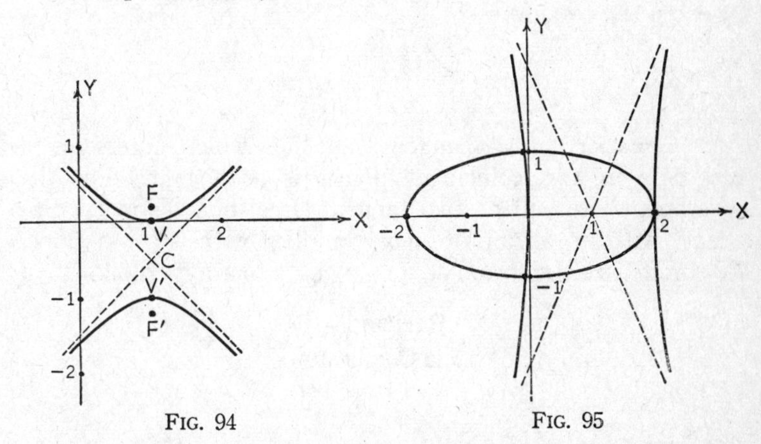

FIG. 94 FIG. 95

Illustration 2. Find the points of intersection of $x^2 + 4\,y^2 - 4 = 0$ and $4\,x^2 - y^2 - 8\,x = 0$ and sketch.

Solution. These curves are an ellipse and a hyperbola respectively. Algebraically it is easy to solve the equations simultaneously for x and y: solve the first equation for y^2 and substitute in the second equation. The points of intersection are $(-\frac{2}{17}, \pm\frac{12}{17}\sqrt{2})$, $(2, 0)$. In standard form the equations are

$$\frac{x^2}{4} + \frac{y^2}{1} = 1 \quad \text{and} \quad \frac{(x-1)^2}{1} - \frac{y^2}{4} = 1.$$

57. Equation of a Hyperbola Referred to Its Asymptotes. In § 6 we remarked that skewed axes were sometimes very useful. If the asymptotes are used as the coordinate axes the equation of the hyperbola takes the form

(1) $$xy = k.$$

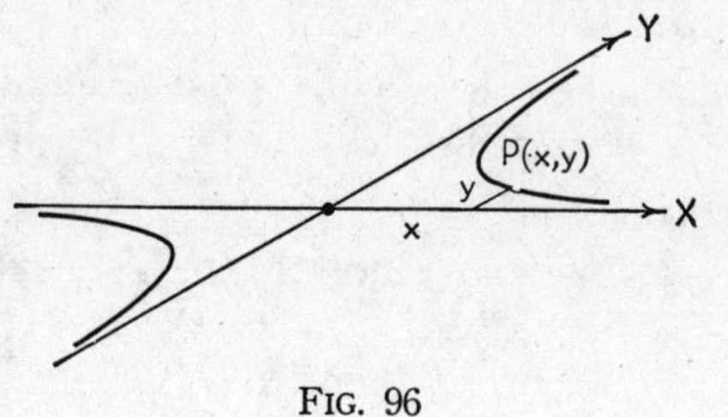

FIG. 96

The standard equation

$$\frac{x^2}{a^2} - \frac{y^2}{b^2} = 1$$

can be written in the form

$$(ay + bx)(ay - bx) + a^2b^2 = 0$$

where the asymptotes are $y = \pm \frac{b}{a}x$. If we call these asymptotes new coordinate axes y' and x' respectively, the hyperbola will have the equation

(2) $$x'y' = \frac{a^2 + b^2}{4}.$$

This is of form (1).

58. Equation of a Tangent. A line which intersects a hyperbola in two coincident points is a tangent. For the hyperbola there will be two tangents [real and distinct, coincident (with an asymptote), or complex] with a given slope. We summarize the formulae for tangents to a hyperbola.

HYPERBOLA
(and Conjugate)

$$\frac{x^2}{a^2} - \frac{y^2}{b^2} = 1$$

$$-\frac{x^2}{a^2} + \frac{y^2}{b^2} = 1$$

$$\frac{(x-h)^2}{a^2} - \frac{(y-k)^2}{b^2} = 1$$

$$-\frac{(x-h)^2}{a^2} + \frac{(y-k)^2}{b^2} = 1$$

TANGENT

At (x_1, y_1)	With Slope m
$\frac{xx_1}{a^2} - \frac{yy_1}{b^2} = 1$	$y = mx \pm \sqrt{a^2m^2 - b^2}$

$$-\frac{xx_1}{a^2}+\frac{yy_1}{b^2}=1 \qquad y=mx\pm\sqrt{b^2-a^2m^2}$$

$$\frac{(x-h)(x_1-h)}{a^2}-\frac{(y-k)(y_1-k)}{b^2}=1 \qquad y-k=m(x-h)\pm\sqrt{a^2m^2-b^2}$$

$$-\frac{(x-h)(x_1-h)}{a^2}+\frac{(y-k)(y_1-k)}{b^2}=1 \qquad y-k=m(x-h)\pm\sqrt{b^2-a^2m^2}$$

Note that for real tangents with slope m the quantity under the radical must be positive. If, for a given slope, the tangents are real for a particular hyperbola, then the tangents are complex for the conjugate hyperbola.

Illustration 1. Write the equation of (a) the tangent and (b) the normal to the hyperbola $16x^2-9y^2-128x-54y+31=0$ at $(\frac{1}{4}, 0)$.

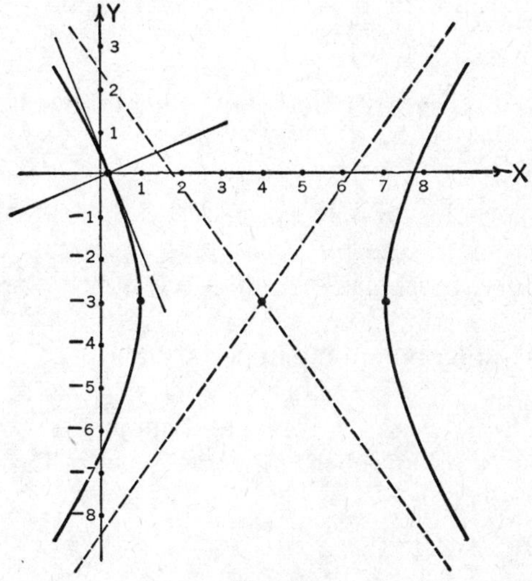

FIG. 97

Solution. (a) The hyperbola in standard form is

$$\frac{(x-4)^2}{9}-\frac{(y+3)^2}{16}=1$$

and the equation of the tangent at $(\frac{1}{4}, 0)$ is

$$\frac{(x-4)(\frac{1}{4}-4)}{9}-\frac{(y+3)(0+3)}{16}=1.$$

This reduces to $20x + 9y - 5 = 0$.

(b) The equation of the normal is

$$y = \tfrac{9}{20}\left(x - \tfrac{1}{4}\right).$$

Illustration 2. Write the equations of the tangents to $2x^2 - y^2 = 1$ with slope 2.

Solution. Here $a^2 = \frac{1}{2}$, $b = 1$, $m = 2$; and the equations of the tangents are

$$y = 2x \pm 1.$$

It is left as an exercise for the student to show that the points of tangency are (1, 1) and (−1, −1).

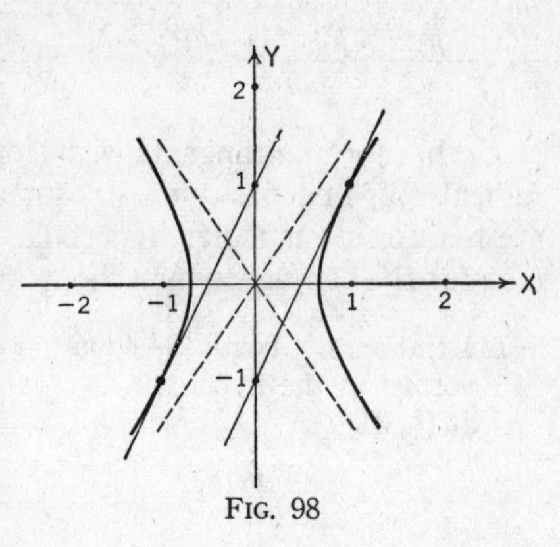

FIG. 98

59. Properties of a Hyperbola. The following properties are listed without proof. They will serve as good exercises for the student.

I. All hyperbolas of like eccentricity are essentially alike and by a proper choice of *scales* (and axes) can be made to coincide. But hyperbolas of unlike eccentricity are unlike in "shape."

II. Let P be any point on a hyperbola and let Q and R be the projections of P onto the transverse and conjugate axes ($2a$ and $2b$) respectively. Then

$$\frac{\overline{PR}^2}{a^2} - \frac{\overline{PQ}^2}{b^2} = 1.$$

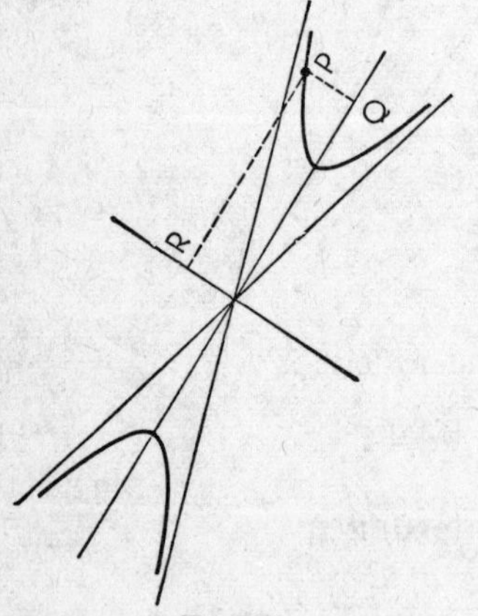

FIG. 99

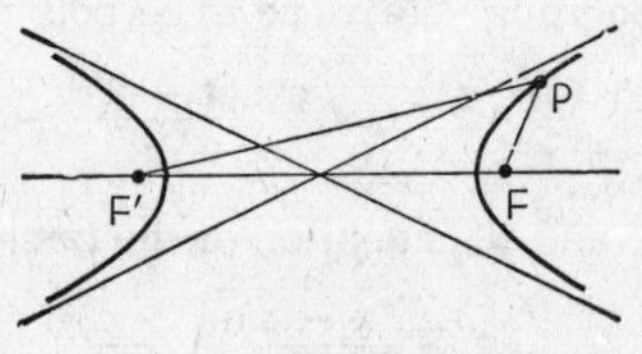

FIG. 100

This fundamental property can be used to write down the standard forms of the equation and can also serve as a definition of a hyperbola. (Fig. 99.)

III. The numerical difference of the focal radii for any point P on a hyperbola is constant and equal to the transverse axis:

$$|PF' - PF| = 2a.$$

This also is a fundamental property and can be used as a definition of the hyperbola. Thus *a hyperbola is the locus of a point the numerical difference of whose distances from two fixed points is a constant.* (Fig. 100.)

IV. A tangent to a hyperbola makes equal angles with the focal radii drawn to the point of tangency. (Fig. 101.)

V. The latus rectum of a hyperbola is a third proportional to the axes.

VI. The product of the perpendiculars on the asymptotes from any point on a hyperbola is a constant.

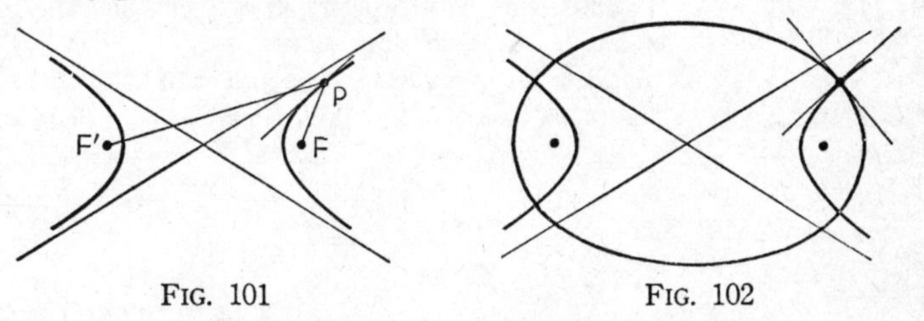

FIG. 101 FIG. 102

VII. Any ellipse and hyperbola which have the same foci intersect at right angles. (Fig. 102.)

EXERCISES

Reduce the following equations to standard form; compute the semitransverse and semiconjugate axes and the eccentricity; write the coordinates of the center, vertices, and foci; write the equations of the directrices and asymptotes; compute the latus rectum; write the equation of the conjugate hyperbola.

1. $x^2 - 2y^2 + 4x + y - \frac{1}{8} = 0.$

Ans. $\frac{(x+2)^2}{4} - \frac{(y-\frac{1}{4})^2}{2} = 1$; $a = 2$, $b = \sqrt{2}$, $ae = \sqrt{6}$, $e = \frac{\sqrt{6}}{2}$;
$C(-2, \frac{1}{4})$, $V(0, \frac{1}{4})$, $V'(-4, \frac{1}{4})$, $F(-2+\sqrt{6}, \frac{1}{4})$, $F'(-2-\sqrt{6}, \frac{1}{4})$;
directrices: $x = -2 \pm \frac{2}{3}\sqrt{6}$, asymptotes: $y - \frac{1}{4} = \pm \frac{\sqrt{2}}{2}(x+2)$;
latus rectum $= 2$; conjugate hyperbola: $-\frac{(x+2)^2}{4} + \frac{(y-\frac{1}{4})^2}{2} = 1.$

2. $4x^2 - y^2 + 2y + 3 = 0$.

Ans. $-\frac{x^2}{1} + \frac{(y-1)^2}{4} = 1$; $a = 2$, $b = 1$, $ae = \sqrt{5}$, $e = \frac{1}{2}\sqrt{5}$; $C(0, 1)$, $V(0, 3)$, $V'(0, -1)$, $F(0, 1 + \sqrt{5})$, $F'(0, 1 - \sqrt{5})$; directrices: $y = 1 \pm \frac{4}{5}\sqrt{5}$, asymptotes: $y - 1 = \pm 2x$; latus rectum $= 1$; conjugate hyperbola: $\frac{x^2}{1} - \frac{(y-1)^2}{4} = 1$.

3. Write the equation of the hyperbola with one vertex at $(5, -4)$ and with asymptotes $y + 4 = \pm \frac{3}{2}(x - 3)$. *Ans.* $\frac{(x-3)^2}{4} - \frac{(y+4)^2}{9} = 1$.

4. Find the equation of $3x^2 - y^2 - 3 = 0$ referred to its asymptotes as axes. *Ans.* $x'y' = 1$.

Find the equation(s) of the tangent(s) to

5. $(x - 1)^2 - y^2 - 1 = 0$ at $(0, 0)$. *Ans.* $x = 0$.

6. $9(x + 1)^2 - 2y^2 + 9 = 0$ with slope $\frac{1}{2}$. *Ans.* $y = \frac{1}{2}(x + 1) \pm \frac{1}{2}\sqrt{17}$.

7. Find the locus of a point which moves so that the numerical difference of its distances from $(-3, 0)$ and $(3, 0)$ is always 4. *Ans.* $\frac{x^2}{4} - \frac{y^2}{5} = 1$.

8. Find the locus of points from which the sound of a gun and the ping of the ball on the target can be heard simultaneously.

Ans. A hyperbola with gun and target at the foci.

9. Show that $x^2 - 2y^2 - 2x - 4y - 1 = 0$ plots two straight lines.

CHAPTER IX

CONIC SECTIONS

60. Sections of a Cone by a Plane. The parabola, ellipse, and hyperbola are all members of a class of curves called *conic sections.* A circle is a special case of an ellipse where the major and minor axes are equal. It was known to the early Greeks that these curves could be obtained by cutting a cone with a plane, and the name derives from this fact.

Consider a right circular cone with vertex A and a tangent sphere S touching the cone along the circle C. Let the plane π tangent to the sphere at F cut the cone in the curve K and the plane γ of the circle C in the line D. Then the curve K is a conic with focus F and directrix D.

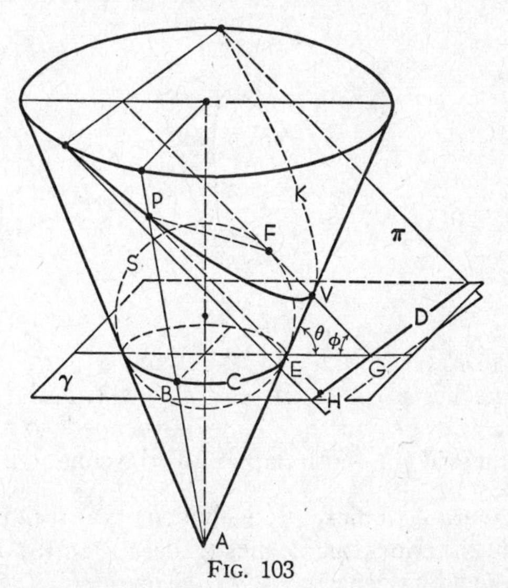

FIG. 103

For consider any element of the cone cutting K at P and C at B. Then $PB = PF$ since any two tangents to a sphere from a point are the same length (solid geometry). Further, every element such as PB makes a constant angle $\theta = GEV$ with plane γ; and a perpendicular PH upon D from P makes a constant angle $\phi = EGV$ with γ. Now from solid geometry it is known that the lengths of any two lines from a point P to a plane γ are inversely proportional to the sines of the angles which the lines make with the plane. Thus

$$(1) \qquad \frac{PF}{PH} = \frac{\sin\phi}{\sin\theta} = \text{constant},$$

and K is a conic with F the focus, D the directrix, and $e = \dfrac{\sin\phi}{\sin\theta}$. The vertex of the conic is V.

If the cutting plane is perpendicular to the axis of the cone it cuts out a circle and $\phi = 0 = e$. Thus a circle is a conic section with eccentricity zero. If the cutting plane is parallel

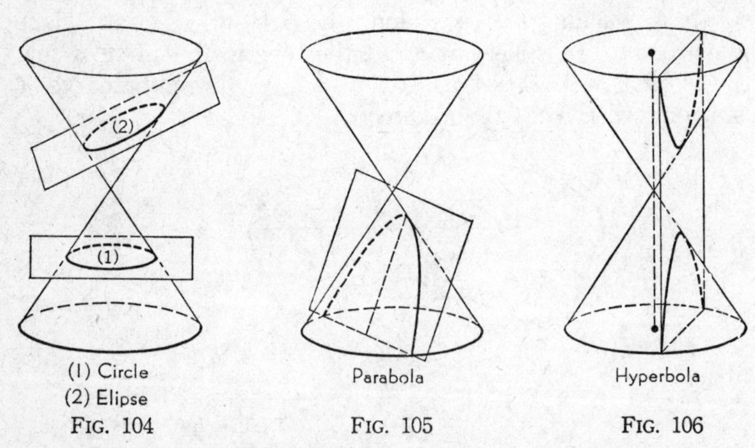

(1) Circle
(2) Elipse
FIG. 104

Parabola
FIG. 105

Hyperbola
FIG. 106

to a generator, $\phi = \theta$ and $e = 1$, giving a parabola. If the cutting plane lies in between, $\phi < \theta$, $e < 1$, and an ellipse is determined. If the cutting plane is inclined so that $\phi > \theta$, $e > 1$, then it will cut (both nappes of) the cone in a hyperbola.

61. Degenerate Conics. It is evident that a plane can also cut a cone in (a) two straight lines (generators), (b) one straight line (two coinciding generators), or (c) one point (vertex). All of these are called *degenerate conics*. And we may think of

(a) As being the limiting case of a hyperbola as the cutting plane moves into a position containing two distinct generators of the cone. An algebraic example is $x^2 - y^2 = 0$, $(AC < 0)$, hyperbola coinciding with its asymptotes.

(b) As being the limiting case of a parabola as the cutting plane moves into coincidence with one generator. An algebraic example is $y^2 = 0$, $(AC = 0)$, parabola coinciding with its axis.

(c) As being the limiting case of an ellipse as the cutting plane moves into a position containing only the vertex. An algebraic example is $x^2 + y^2 = 0$, $(AC > 0)$, ellipse coinciding with its center.

Analytically there is another degenerate conic, namely a pair of parallel lines. An example is $y(y - 1) \equiv y^2 - y = 0$, $(AC = 0)$, parabola. This cannot be gotten as a section of a cone. However, it can be obtained by cutting a cylinder (limiting case of a cone as the vertex recedes to infinity) with a plane parallel to the axis.

CHAPTER X

TRANSFORMATIONS OF COORDINATES

62. Transformations. A transformation is essentially a substitution of a function of one or more variables for a given variable. Thus the substitution of $x = 5x'$ in $y = f(x)$ yields $y = f(5x')$ and this in turn says that y is some (other) function of x', say $y = F(x')$. A transformation of this character is a scale transformation since the scale (in the X-direction) only is affected. (See Property I for parabola, ellipse, hyperbola, §§ 45, 51, 59 respectively.)

There are many types of transformations in mathematics and we now study two particular ones important in analytic geometry. By making use of one or both of these it often happens that the equation of a graph can be simplified or some special property of the locus can be prominently displayed.

63. Translation. About the simplest of transformations is a *translation* of the axes parallel to themselves.

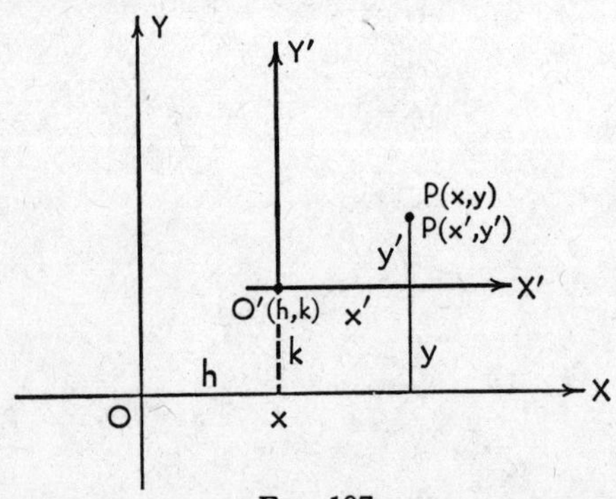

FIG. 107

Let OX, OY be a set of rectangular axes and $O'X'$, $O'Y'$ another set, $O'X'$ parallel to OX and $O'Y'$ parallel to OY, the directions remaining the same. (Fig. 107.) Then, if with respect to these axes a point P has coordinates $P(x, y)$ and $P(x', y')$, it is evident that

$$x = x' + h, \quad y = y' + k, \tag{1}$$

where (h, k) are the coordinates, with respect to OX and OY, of the new origin O'.

Equations (1) are the *equations of the transformation.* By making these substitutions in a given equation a new equation of the same graph is obtained, referred now to the new (translated) axes.

Illustration 1. Translate axes to the new origin (2, 3) and thus reduce the equation of the curve $x^2 - 4y^2 - 4x + 24y - 36 = 0$ to a new equation.

Solution. Here we must substitute

$$x = x' + 2, \quad y = y' + 3$$

in the given equation. This yields

$$(x' + 2)^2 - 4(y' + 3)^2 - 4(x' + 2) + 24(y' + 3) - 36 = 0,$$

which reduces to

$$\frac{x'^2}{4} - \frac{y'^2}{1} = 1, \tag{1}$$

a simpler equation. The process, here, is equivalent to completing the square. For the original equation, in standard form, is found, by this process, to be

$$\frac{(x-2)^2}{4} - \frac{(y-3)^2}{1} = 1 \text{ (a hyperbola).} \tag{2}$$

Setting $x = x' + 2$ and $y = y' + 3$ in (2) will obviously reduce this equation to form (1).

A translation can always be found which will remove the first-degree terms from the equation of a (non-degenerate) conic section.

Illustration 2. By a translation transform the equation $x^2 - 2xy - 5y^2 + x - 3y = 0$ into one in which there are no first-degree terms.

Solution. Substitute $x = x' + h$, $y = y' + k$ into the equation and hence determine what h and k must be in order that the new equation have no first-degree terms. The transformed equation is

$$(x'+h)^2-2(x'+h)(y'+k)-5(y'+k)^2+(x'+h)-3(y'+k)=0,$$

and this reduces to

$$(1)\quad x'^2-2\,x'y'-5\,y'^2+(2\,h-2\,k+1)x'+(-2\,h-10\,k-3)y' \\ +h^2-2\,hk-5\,k^2+h-3\,k=0.$$

We must set the coefficients of x' and y' equal to zero. Thus

$$(2)\qquad \begin{aligned} 2\,h-2\,k+1&=0,\\ -2\,h-10\,k-3&=0.\end{aligned}$$

Solving these simultaneously we get

$$h=-\tfrac{2}{3},$$
$$k=-\tfrac{1}{6};$$

and with these values of h and k the transformation reduces the equation to

$$x'^2-2\,x'y'-5\,y'^2-\tfrac{1}{12}=0.$$

Note that while the linear terms have been removed the *same* quadratic terms are present as before. A translation will not remove terms of the second degree from an equation of second degree. Indeed a translation always leaves unchanged the terms of highest degree in any equation thus transformed.

64. Rotation. Let a set of rectangular axes OX and OY, as a rigid system, be rotated counterclockwise about O through a positive angle θ into a new position OX' and OY'. The substitutions describing such a transformation are called the formulae for the *rotation of axes*. We consider only the case where $\theta < 90°$, but the final results hold for any angle.

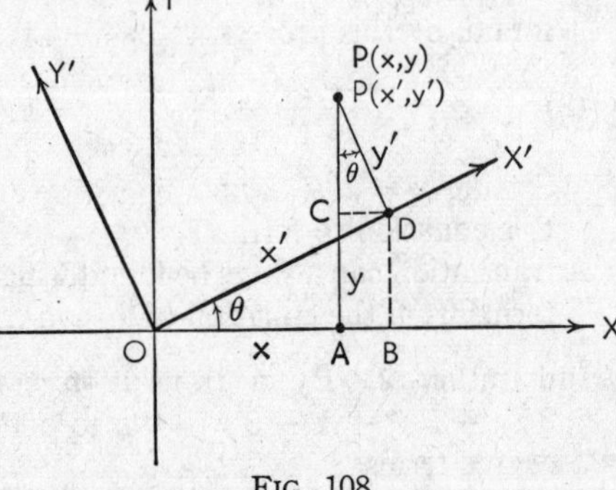

FIG. 108

From Fig. 108 we see that

$$OA = x = OB - AB$$
$$= OB - CD,$$

and

$$AP = y = AC + CP$$
$$= BD + CP.$$

Further, from trigonometry, $OB = x' \cos\theta$, $CD = y' \sin\theta$, $BD = x' \sin\theta$, and $CP = y' \cos\theta$. Hence

(1)
$$x = x' \cos\theta - y' \sin\theta,$$
$$y = x' \sin\theta + y' \cos\theta.$$

Equations (1) are the *equations of the transformation.* By rotating axes the equation of a graph may be simplified.

Illustration. For the equation $x^2 + 4xy + y^2 - 1 = 0$ rotate axes through $\theta = 45°$.

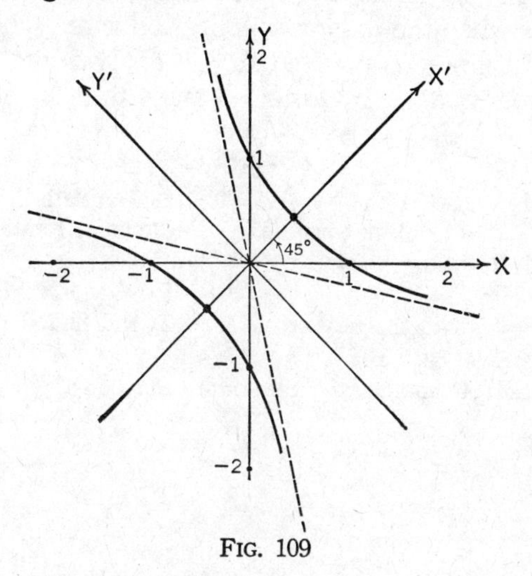

FIG. 109

Solution. The substitutions reduce to

$$x = \frac{\sqrt{2}}{2}(x' - y'),$$
$$y = \frac{\sqrt{2}}{2}(x' + y'),$$

since $\sin 45° = \cos 45° = \sqrt{2}/2$.

The transformed equation is

$$\tfrac{1}{2}(x'^2 - 2x'y' + y'^2) + 2(x'^2 - y'^2) + \tfrac{1}{2}(x'^2 + 2x'y' + y'^2) - 1 = 0,$$

which reduces to

$$3x'^2 - y'^2 - 1 = 0.$$

This is not only a simpler equation (the xy-term has been removed) but it is essentially in standard form and from it the graph (hyperbola) can be constructed with ease. It would be much more difficult to plot the curve with respect to the original axes.

By combining translations and rotations a curve may be referred to axes in any direction with any point in the plane as origin. These transformations will change the equation of the curve, but the curve itself will not be altered in any way — it will only be moved from one position to another. Translations and rotations do not affect the degree of the equation to which they are applied. They do not in general give the same results when the order of application is interchanged.

The next chapter is devoted to the analysis of the general second-degree equation by means of these transformations.

EXERCISES

1. Given the equation $2x - y + 2 = 0$. (a) Translate the axes to the new origin $(0, 2)$. (b) For the equation resulting from (a) rotate axes through an angle θ such that $\tan\theta = 2$, i.e., $\theta = \tan^{-1} 2$.

Ans. (a) $2x' - y' = 0$; (b) $y'' = 0$.

2. Translate axes to the new origin $(1, -2)$ and find the transformed equation of $x^3 - 3x^2 - y^2 + 3x - 4y - 5 = 0$. *Ans.* $y'^2 = x'^3$.

3. Rotate axes 90° and find the transformed equation of $y^2 = 4px$.

Ans. $x'^2 = -4py'$.

CHAPTER XI

GENERAL EQUATION OF SECOND DEGREE

65. Classification of Conics. We have seen that the general equation of a conic section is of the form

$$(1) \qquad Ax^2 + Bxy + Cy^2 + Dx + Ey + F = 0.$$

Conversely any equation of the form (1) represents a conic, real or complex, proper or degenerate. To distinguish four cases we first define

$$\Delta = \begin{vmatrix} 2A & B & D \\ B & 2C & E \\ D & E & 2F \end{vmatrix}$$

as the *discriminant* of equation (1). The discriminant plays an important role in the classification of conics, which is outlined as follows.

Outline
Classification of Conics

Case	Conditions	Type of Locus
I	$B^2 - 4AC = 0$ and $\Delta = 0$	Two parallel or coincident lines. (Degenerate parabola.)
II	$B^2 - 4AC \neq 0$ and $\Delta = 0$	Two intersecting lines or one point. (Degenerate central conic.)
III	$B^2 - 4AC = 0$ and $\Delta \neq 0$	A parabola.
IV	$B^2 - 4AC \neq 0$ and $\Delta \neq 0$	A central conic: An ellipse (real or complex) if $B^2 - 4AC < 0$; a hyperbola if $B^2 - 4AC > 0$.

The conic is degenerate if and only if $\Delta = 0$; it is proper if and only if $\Delta \neq 0$. We may combine cases I and III and cases II and IV by saying equation (1) will represent

(a) A parabola, two parallel lines, or two coincident lines if the terms of the second degree form a perfect square;

(b) An ellipse, a point, or no locus if the terms of the second degree have conjugate complex factors;

(c) A hyperbola or two intersecting lines if the factors of the second-degree terms are real and distinct.

We have also noted that the axis of the parabola and the axes of the ellipse and hyperbola are *not* parallel to a coordinate axis if and only if the xy-term is present.

66. Removal of the xy-Term. *A properly chosen rotation of axes will always remove the xy-term from an equation of the second degree.* For if we apply the rotation formulae to the general equation we get as the coefficient of the $x'y'$-term

$$-(A - C)\, 2 \sin \theta \cos \theta + B(\cos^2 \theta - \sin^2 \theta).$$

If we equate this to zero and make use of the trigonometric identities $2 \sin \theta \cos \theta = \sin 2\theta$, $\cos^2 \theta - \sin^2 \theta = \cos 2\theta$, the result is

$$\tan 2\theta = \frac{B}{A - C}. \tag{1}$$

An angle θ, between 0° and 90°, is always determined by equation (1) for any values of A, B, and C. A rotation of axes through this angle θ, therefore, will produce an equation of the conic referred to OX' and OY' in which there is no $x'y'$-term. Obviously the primes could be suppressed from the final equation; they are retained only to show clearly that the new axes have been derived from old ones.

When $\tan 2\theta = \dfrac{B}{A - C}$ it results from trigonometry that

$$\sin \theta = \sqrt{\frac{1 - \cos 2\theta}{2}}, \qquad \cos \theta = \sqrt{\frac{1 + \cos 2\theta}{2}}. \tag{2}$$

We need these for the equations of the transformation.

67. Reduction of the General Equation to Standard Form. Theoretically the reduction of the general equation of the second degree to standard form can be accomplished by first rotating axes to remove the xy-term and second completing the square on the transformed equation. Practically this leads to computational difficulties in most cases because of the irrationalities introduced by the square roots in the formulae for $\sin\theta$ and $\cos\theta$ in (2), § 66 above.

In order to illustrate the theory without undue numerical complications the following example has been prepared so as to avoid the irrationalities.

Illustration. Reduce the conic $8x^2+24xy+y^2+x+2y+1=0$ to standard form by rotating axes and completing the square.

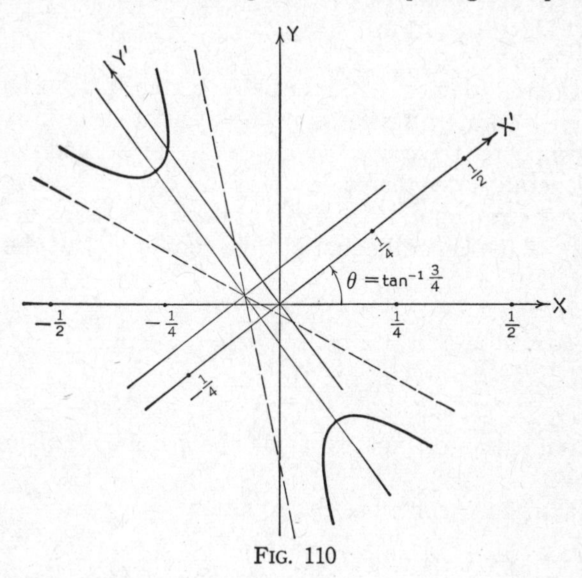

FIG. 110

Solution.

$$\tan 2\theta = \tfrac{24}{7},$$
$$\cos 2\theta = \tfrac{7}{25},$$
$$\sin\theta = \tfrac{3}{5},$$
$$\cos\theta = \tfrac{4}{5}.$$

The formulae of rotation are therefore

$$x = \tfrac{1}{5}(4x' - 3y'),$$
$$y = \tfrac{1}{5}(3x' + 4y'),$$

and these reduce the given equation to

$$17x'^2 - 8y'^2 + 2x' + y' + 1 = 0.$$

On completing the square we get, as the standard form of the conic (a hyperbola),

$$-\frac{(x'+\frac{1}{17})^2}{\frac{(23)(23)}{(17)(17)(32)}}+\frac{(y'-\frac{1}{16})^2}{\frac{(23)(23)}{(17)(16)(16)}}=1.$$

Even with such numbers as these for a and b it is not difficult to plot the hyperbola.

EXERCISES

Remove the xy-term, complete the square, and trace the curve.

1. $9x^2+24xy+16y^2+90x-130y=0.$ *Ans.* $(x'-1)^2=6(y'+\frac{1}{6}).$

2. $5x^2+8xy+5y^2+\sqrt{2}\,x-\sqrt{2}\,y=0.$ *Ans.* $\frac{x^2}{\frac{1}{9}}+\frac{(y-1)^2}{1}=1.$

68. Central Conics. It is usually better to translate axes first and to rotate afterwards. The translation is to be made to the center for the *central conics*, circle, ellipse, and hyperbola (or degenerate cases thereof).

For a central conic $B^2-4AC\neq 0$ and, in terms of the coefficients of the general equation, the coordinates of the center are

$$(1)\qquad h=\frac{2CD-BE}{B^2-4AC},\qquad k=\frac{2AE-BD}{B^2-4AC}.$$

Note that these do not exist for the parabola, where $B^2-4AC=0$.

The translated conic has the equation

$$(2)\qquad Ax'^2+Bx'y'+Cy'^2-\frac{1}{2}\cdot\frac{\Delta}{B^2-4AC}=0.$$

Rotation is now performed on equation (2), thus completing the reduction.

Illustration. Reduce the equation $x^2+xy+y^2-3x-1=0$ to standard form by first translating axes to the center and then rotating.

Solution. We compute

$$B^2-4AC=-3 \text{ (ellipse)},$$

$$\Delta=\begin{vmatrix} 2 & 1 & -3\\ 1 & 2 & 0\\ -3 & 0 & -2\end{vmatrix}=-24,$$

$$h=2,$$

$$k=-1.$$

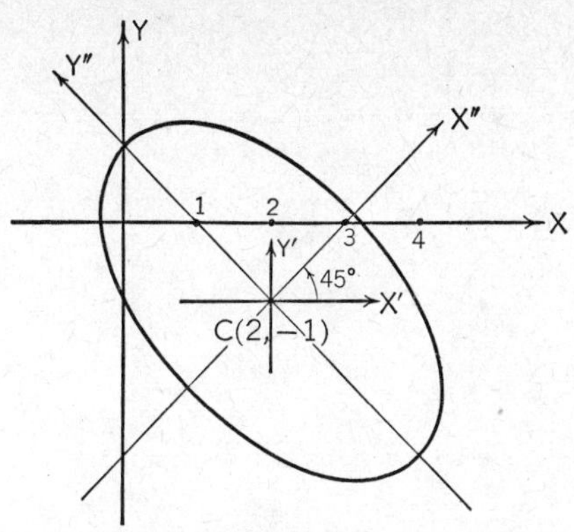

FIG. 111

Making use of equation (2) the equation of the conic referred to axes parallel to the original axes and through the center of the ellipse is

$$x'^2 + x'y' + y'^2 - 4 = 0.$$

Now

$$\tan 2\,\theta = \frac{1}{1-1}$$

which tells us that $2\,\theta = 90°$ and $\theta = 45°$.

The formulae for rotation are

$$x' = \frac{1}{\sqrt{2}}\,(x'' - y''),$$

$$y' = \frac{1}{\sqrt{2}}\,(x'' + y'')$$

and these reduce the equation to $3\,x''^2 + y''^2 = 8$, or, in standard form, to

$$\frac{x''^2}{\frac{8}{3}} + \frac{y''^2}{8} = 1.$$

EXERCISES

Translate axes to the center, then rotate these into coincidence with the axes of the conic, and sketch.

1. $19\,x^2 + 4\,xy + 16\,y^2 - 212\,x + 104\,y - 356 = 0.$

Ans. $19\,x'^2 + 4\,x'y' + 16\,y'^2 - 1200 = 0,\ \frac{x''^2}{60} + \frac{y''^2}{80} = 1.$

2. $3\,x^2 + 12\,xy - 2\,y^2 + 18\,x + 8\,y + 12 = 0.$

Ans. $3\,x'^2 + 12\,x'y' - 2\,y'^2 - 1 = 0,\ \frac{x''^2}{\frac{1}{7}} - \frac{y''^2}{\frac{1}{6}} = 1.$

69. The Parabola. The general parabola $Ax^2 + Bxy + Cy^2 + Dx + Ey + F = 0$, $B^2 - 4\,AC = 0$, can best be treated by computing the coordinates of the focus and the equation of the directrix, which are given by

$$\text{Focus:}\quad \left(\frac{2\,C'E' + D'(A' - B')}{2(C'^2 + D'^2)},\ \frac{2\,D'E' - C'(A' - B')}{2(C'^2 + D'^2)}\right)$$

$$\text{Directrix:}\quad 2\,D'x + 2\,C'y - A' - B' = 0,$$

$$\text{where } A' = CF - \frac{E^2}{4},\ B' = AF - \frac{D^2}{4},\ C' = \frac{BD - 2\,AE}{4},$$

$$D' = \frac{BE - 2\,CD}{4},\quad E' = \frac{DE - 2\,BF}{4}.$$

These formulae seem a little forbidding but with the information they yield the vertex and the axis can be determined and the conic readily plotted. Note that this method involves neither translation nor rotation of axes.

Illustration. Sketch the conic $x^2+2\,xy+y^2+2\,x-2\,y+4=0$.

Solution. Since $B^2 - 4\,AC = 0$ the conic is a parabola; $\Delta = -32$ and the parabola is proper. We compute $A' = 3$, $B' = 3$, $C' = 2$, $D' = -2$, $E' = -5$. The coordinates of the focus turn out to be $(-\frac{5}{4}, \frac{5}{4})$ and the equation of the directrix is $2\,x - 2\,y + 3 = 0$. The axis passes through the focus and is perpendicular to the directrix; hence its equation is $y - \frac{5}{4} = -(x + \frac{5}{4})$, which reduces to $y = -x$. The directrix and the axis intersect at $(-\frac{3}{4}, \frac{3}{4})$ and the vertex is halfway along the axis toward the focus from this point. The vertex is therefore at $(-1, 1)$.

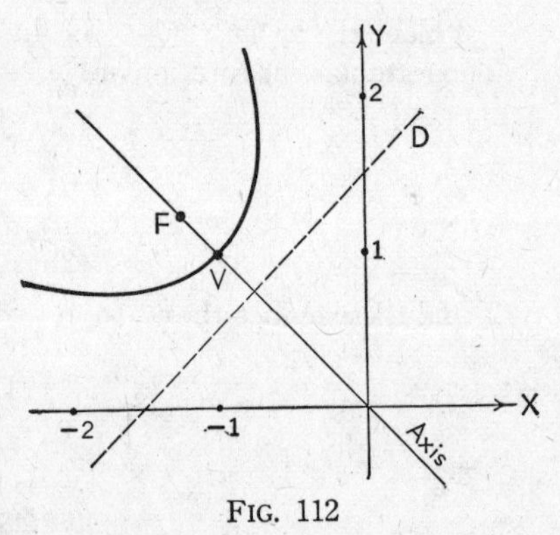

FIG. 112

EXERCISES

1. Sketch the parabola $x^2 + 2\,xy + y^2 - 2\,x + 6\,y - 2 = 0$.

Show: $F(-\frac{3}{8}, -\frac{5}{8})$; equation of directrix, $4\,x - 4\,y + 7 = 0$; equation of axis, $x + y + 1 = 0$; $V(-\frac{7}{8}, -\frac{1}{8})$.

2. Sketch the parabola $x^2 - 4\,xy + 4\,y^2 + x - y - 1 = 0$.

Show: $F(\frac{21}{10}, \frac{6}{5})$; equation of directrix, $4x + 2y - 11 = 0$; equation of axis, $10x - 20y + 3 = 0$; $V(\frac{53}{25}, \frac{121}{100})$.

70. Invariants. Often the general equation of the conic is written in the form

$$ax^2 + 2hxy + by^2 + 2gx + 2fy + c = 0 \tag{1}$$

where the factor 2 is inserted in the xy-, x-, and y-terms for later simplification in formulae. For this form of the equation we define

$$I = a + b,$$
$$\overline{C} = ab - h^2,$$
$$J = ab + bc + ca - f^2 - g^2 - h^2,$$
$$\overline{D} = \begin{vmatrix} a & h & g \\ h & b & f \\ g & f & c \end{vmatrix}.$$

(There should be no confusion between this "$\overline{C}$" and "$\overline{D}$" and the C and D we have been using as coefficients in the general equation.)

Each of these four quantities is *invariant* under rotation of axes; that is, they are equal respectively to the corresponding quantities after a rotation is performed. (I, $\overline{C}$, $\overline{D}$, and the *sign* of J are also invariant under translation.) They are useful in the classification of conics as in the following table.

Detail
Classification of Conics

Case	Conditions on the Invariants	Type of Locus
Proper Conic; $\overline{D} \neq 0$	$\overline{C} > 0$; $I, \overline{D}$ opposite in sign, $a = b, h = 0$	Circle
	$\overline{C} > 0$; $I, \overline{D}$ opposite in sign	Ellipse
	$\overline{C} < 0$	Hyperbola
	$\overline{C} = 0$	Parabola
	$\overline{C} > 0$; $I, \overline{D}$ same in sign	No real locus
Degenerate Conic; $\overline{D} = 0$	$\overline{C} < 0$	Two intersecting lines
	$\overline{C} = 0, J < 0$	Two parallel lines
	$\overline{C} = 0, J = 0$	Two coincident lines
	$\overline{C} > 0$	A point
	$\overline{C} = 0, J > 0$	No real locus

The student will find it worth while to prepare a similar table for the equation $Ax^2 + Bxy + Cy^2 + Dx + Ey + F = 0$, using the two invariants $B^2 - 4AC$ and Δ as already defined (but simply related to invariants $\overline{C}$ and $\overline{D}$ of this paragraph).

71. Systems of Conics. Two conics intersect in four points (4 real, or 2 real and 2 complex, or 4 complex) since each equation is of the second degree. If $U = 0$ and $V = 0$ represent two conics, then $U + kV = 0$, for any constant k, is a conic through the points of intersection of $U = 0$ and $V = 0$. (See Systems of Lines, § 26; Systems of Circles, § 37.) One condition placed on the conic $U + kV = 0$ will determine k.

Illustration 1. Find the conic which passes through the intersections of the ellipse $x^2 - xy + y^2 - x + 1 = 0$ and the hyperbola $x^2 + xy - y^2 + y - 4 = 0$ and also through (1, 2).

Solution. The equation is of the form

$$(x^2 - xy + y^2 - x + 1) + k(x^2 + xy - y^2 + y - 4) = 0,$$

and this must be satisfied by (1, 2). Hence $k = 1$ and the final equation of the conic sought is

$$2x^2 - x + y - 3 = 0 \text{ (parabola)}.$$

If $U = 0$ is a conic and $L = 0$ is a straight line, then $U + kL^2 = 0$ represents a conic tangent to $U = 0$ at the points of intersection of $U = 0$ and $L = 0$. For we may think of $L^2 = 0$ as a (degenerate) conic (two coincident lines) intersecting $U = 0$ in two pairs of two coincident points each.

Illustration 2. Find the conic tangent to $U \equiv x^2 - y^2 - y + 1 = 0$ at the points of intersection of $U = 0$ and $L \equiv 3x - 2y - 1 = 0$ and passing through $(-1, 0)$.

Solution. The equation is of the form

$$(x^2 - y^2 - y + 1) + k(3x - 2y - 1)^2 = 0.$$

Substituting $(-1, 0)$ into this we get $k = -\frac{1}{8}$, which yields as the equation of the conic sought

$$x^2 - 12xy + 12y^2 - 6x + 12y - 7 = 0.$$

(See the Illustration in § 73.)

72. Conic through Five Points. Since the general equation of the second degree has five effective coefficients in it there is one and only one conic passing through five arbitrary points (no three lying on the same straight line).

Illustration. Find the conic passing through (1, 1), (2, 1), (3, −1), (−3, 2), and (−2, −1).

Solution. By substituting these points into the general equation $Ax^2 + Bxy + Cy^2 + Dx + Ey + F = 0$ we obtain five simultaneous equations.

$$\begin{aligned} A + B + C + D + E + F &= 0, \\ 4A + 2B + C + 2D + E + F &= 0, \\ 9A - 3B + C + 3D - E + F &= 0, \\ 9A - 6B + 4C - 3D + 2E + F &= 0, \\ 4A + 2B + C - 2D - E + F &= 0. \end{aligned}$$

Under any circumstances the solution of this system is tedious; but by setting $A = 1$ and using determinants we get: $A = 1$, $B = -1$, $C = -9$, $D = -2$, $E = 4$, and $F = 7$. The conic is, therefore,

$$x^2 - xy - 9y^2 - 2x + 4y + 7 = 0.$$

It is better, however, to make use of the theory of systems of conics as follows. The pair of lines through, say (1, 1), (2, 1) and (3, −1), (−2, −1), may be thought of as a degenerate conic; its equation is the product of the equations of the two lines or

(1) $$(y - 1)(y + 1) = 0.$$

Similarly

(2) $$(2x - 3y + 1)(2x + y - 5) = 0$$

is the equation of the (degenerate) conic consisting of the two lines joining (1, 1), (−2, −1) and (2, 1), (3, −1). These conics intersect in the four points (1, 1), (2, 1), (3, −1), and (−2, −1). Hence

$$(y - 1)(y + 1) + k(2x - 3y + 1)(2x + y - 5) = 0$$

is a conic through these four points. We can determine k so that it passes through the fifth point (−3, 2). We get $k = -\frac{1}{33}$ and the conic is again

$$x^2 - xy - 9y^2 - 2x + 4y + 7 = 0.$$

73. Equation of a Tangent. The equation of the tangent to $Ax^2 + Bxy + Cy^2 + Dx + Ey + F = 0$ at the point (x_1, y_1) is

(1) $$2Axx_1 + B(xy_1 + x_1y) + 2Cyy_1 + D(x + x_1) + E(y + y_1) + 2F = 0.$$

In the notation of §70 the equations of the tangents with slope m are

$$(2)\quad y = mx + \frac{P - mQ \pm \sqrt{(-\overline{D})(bm^2 + 2hm + a)}}{\overline{C}}, \quad \overline{C} \neq 0,$$

where $P = gh - af$ and $Q = fh - bg$.

Illustration. Find the equations of the tangents to $x^2 - y^2 - y + 1 = 0$ and $x^2 - 12xy + 12y^2 - 6x + 12y - 7 = 0$ at the point $(1, 1)$.

Solution. By (1) the tangents are respectively

$$2x - 2y - (y + 1) + 2 = 0, \quad \text{or} \quad 2x - 3y + 1 = 0,$$

and

$$2x - 12(x + y) + 24y - 6(x + 1) + 12(y + 1) - 14 = 0,$$

or

$$2x - 3y + 1 = 0.$$

Since the equations of the tangents are the same the two conics are tangent to each other at the point $(1, 1)$. They are also tangent to each other at $(-1, -2)$. (See Illustration 2, § 71.)

EXERCISES

1. Find the conic passing through the intersections of $x^2 - y^2 = 1$ and $x^2 + y^2 = 4$ and the point $(0, 3)$. *Ans.* $3x^2 + y^2 - 9 = 0$.

2. Find the conic through the five points $(0, 0)$, $(1, 0)$, $(2, 1)$, $(-1, 2)$, and $(-1, -1)$. *Ans.* $3x^2 - 2xy - 3y^2 - 3x + y = 0$.

3. Find the equations of the tangents with slope 1 to $x^2+xy+y^2-3x-1=0$.
Ans. $C' = -\frac{3}{4}$, $D' = \frac{3}{2}$, $\overline{C} = \frac{3}{4}$, $\overline{D} = -3$; the equations of the tangents are $x - y + 1 = 0$ and $x - y - 7 = 0$.

CHAPTER XII

POLES AND POLARS

74. Definitions and Theorems. If $P(x_1, y_1)$ is on the conic

$$Ax^2 + Bxy + Cy^2 + Dx + Ey + F = 0 \tag{1}$$

the equation of the tangent at P is

$$2\,Axx_1 + B(xy_1 + x_1y) + 2\,Cyy_1 + D(x + x_1) + E(y + y_1) + 2\,F = 0. \tag{2}$$

Whether P is on the conic or not, equation (2) represents a line called the *polar* of the *pole* P.

Without proof we list the following theorems concerning poles and polars.

I. If the polar of P_1 passes through P_2, then the polar of P_2 passes through P_1.

II. If the polars of P_1 and P_2 intersect at P, then P is the pole of P_1P_2.

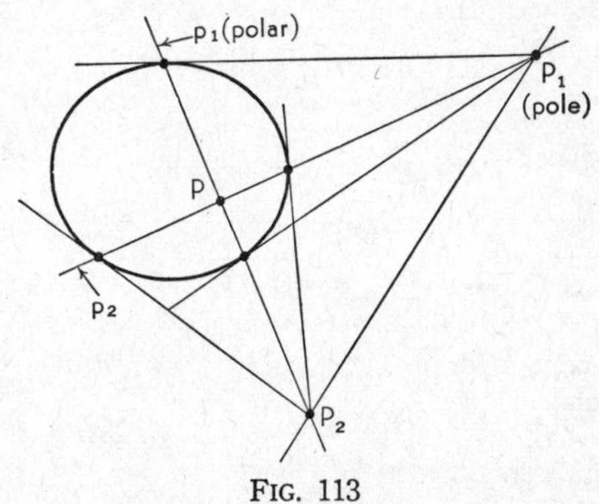

FIG. 113

III. The polar of an exterior point P_1 is the line joining the points of contact of the tangents drawn from P_1.

IV. The polar of an interior point P is the locus of the point of intersection of the tangents at the extremities of every chord through P.

V. The polar of a focus is the corresponding directrix.

VI. There is no (finite) polar of the center of a conic.

VII. Let P be the pole of p and let a secant through P intersect the conic in Q and R and p in S. Then

$$\frac{PQ}{PR} = -\frac{SQ}{SR}.$$

QR is said to be divided *harmonically* by P and S.

Illustration 1. Find the polar of (3, −2) with respect to the conic $xy - y^2 - x + 1 = 0$.

Solution. Substituting (3, −2) into the equation (2) of the polar we get, upon simplifying,

$$3x - 7y + 1 = 0.$$

Illustration 2. Find the pole of $x - 3y + 4 = 0$ with respect to $x^2 - xy + x + y = 0$.

Solution. Let (x_1, y_1) be the coordinates of the pole. Then

$$(2x_1 - y_1 + 1)x - (x_1 - 1)y + (x_1 + y_1) = 0$$

is the equation of the polar and must be the same as $x - 3y + 4 = 0$. Hence the coefficients must be proportional; that is,

$$\frac{2x_1 - y_1 + 1}{1} = \frac{x_1 - 1}{3} = \frac{x_1 + y_1}{4}.$$

Whence

$$5x_1 - 3y_1 = -4,$$
$$x_1 - 3y_1 = 4,$$

which yields $x_1 = -2$, $y_1 = -2$ as the coordinates of the pole.

EXERCISES

1. With respect to $3x^2 - y^2 - x + 2 = 0$ find (a) the polar of $(-1, 6)$, (b) the pole of $x + y - 1 = 0$. *Ans.* (a) $7x + 12y - 5 = 0$, (b) $(-\frac{3}{5}, \frac{23}{10})$.

2. Verify theorem I for the conic $xy - y^2 + x = 0$ and the two points (poles) $P_1(1, 2)$ and $P_2(2, \frac{7}{3})$.

Ans. $p_1 \equiv 3x - 3y + 1 = 0$, which passes through P_2; $p_2 \equiv 5x - 4y + 3 = 0$, which passes through P_1.

3. Find the point of intersection of the tangents to $xy - x - y - 2 = 0$ at the points where it is cut by $3x - y + 1 = 0$. *Ans.* $(3, -5)$.

CHAPTER XIII

DIAMETERS

75. Definitions and Theorems. The locus of the midpoints of a system of parallel chords of a conic is called a *diameter* of the conic. A diameter may be thought of as either a segment of a line or an entire line. Two diameters are said to be *conjugate* when each bisects the chords parallel to the other.

The following theorems apply to the central conics $\frac{x^2}{a^2} \pm \frac{y^2}{b^2} = 1$ (throughout this section read top sign for ellipse and bottom sign for hyperbola). (*Note:* If "a" and "b" are interchanged in the formulae of the ellipse and hyperbola they must be interchanged in the following formulae. See *Notes*, §§ 48 and 54.)

I. Every diameter is a straight line through the center and conversely.

II. The tangent at an end of a diameter is parallel to the system of chords defining that diameter.

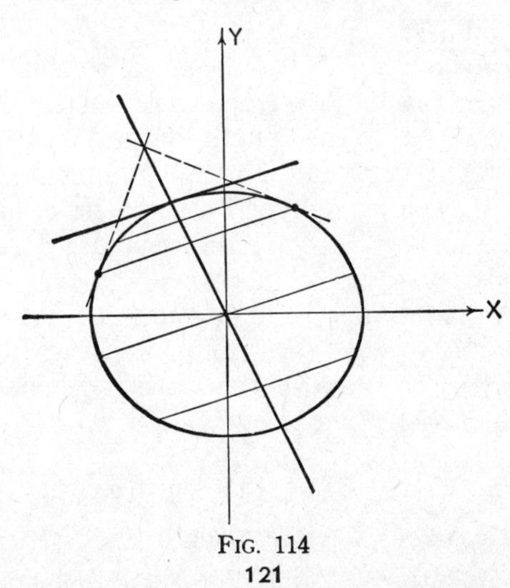

FIG. 114

III. Tangents at the ends of any chord intersect on the diameter bisecting that chord.

IV. The polar of any point on a diameter is parallel to the system of chords defining that diameter.

V. If $y = mx$ is a diameter, the equation of the conjugate diameter is $y = \mp \frac{b^2}{a^2m} x$. Or again, the lines $y = mx$ and $y = m'x$ are conjugate diameters if and only if $mm' = \mp \frac{b^2}{a^2}$.

VI. If $P_1(x_1, y_1)$ is on the conic, the equation of the diameter conjugate to OP_1 is $\frac{xx_1}{a^2} \pm \frac{yy_1}{b^2} = 0$.

VII. If a_1 and b_1 are the lengths of the semiconjugate diameters, then $a_1^2 \pm b_1^2 = a^2 \pm b^2$, a constant. Moreover the angle θ between the conjugate diameters is given by $\sin \theta = ab/a_1b_1$.

VIII. If P_1P_2 and Q_1Q_2 are conjugate diameters, then the parallelogram formed by the tangents to the conic at P_1, P_2, Q_1, Q_2 is of constant area $4\,ab$.

IX. The equations of the ellipse and hyperbola referred to a pair of conjugate diameters as oblique axes are respectively

$$\frac{x^2}{a_1^2} \pm \frac{y^2}{b_1^2} = 1.$$

(See §§ 6 and 57.)

Note on the hyperbola: If one of a pair of conjugate diameters meets the hyperbola the other does not; instead it meets the conjugate hyperbola. The length of a diameter is to be measured therefore between the points of intersection with the hyperbola or the conjugate hyperbola, whichever applies.

For the parabola $y^2 = 4\,px$, theorems I–IV and the two following apply.

X. A parabola has no system of conjugate diameters.

XI. If $y = mx + k$ is a system of chords of slope m, then $y = 2\,p/m$ is the equation of the associated diameter. (For $x^2 = 4\,py$ the diameter is $x = 2\,pm$.)

Illustration 1. For $\frac{x^2}{16} + \frac{y^2}{9} = 1$ the length of a diameter is $a_1 = 5$. Find the length on the conjugate diameter b_1.

Solution. By theorem VII we have

$$a_1^2 + b_1^2 = a^2 + b^2 = 25,$$
$$b_1^2 = 25 - \tfrac{25}{4} = \tfrac{75}{4},$$

or

$$b_1 = \tfrac{5}{2}\sqrt{3}.$$

Illustration 2. In the hyperbola $4x^2 - y^2 = 4$ the slope of a diameter is 1. Find the extremities of the conjugate diameter.

Solution. The slope m' of the conjugate diameter is given by (theorem V)

$$mm' = b^2/a^2 = 4.$$

Therefore

$$m' = 4.$$

The equation of the conjugate diameter is

$$y = 4x$$

and this intersects the conjugate hyperbola $-4x^2 + y^2 = 4$ in the points $\left(\dfrac{\sqrt{3}}{3}, \tfrac{4}{3}\sqrt{3}\right)$ and $\left(-\dfrac{\sqrt{3}}{3}, -\tfrac{4}{3}\sqrt{3}\right)$.

Illustration 3. For the parabola $y^2 = 6x$ a diameter meets the curve in the point for which $y = -2$. Find the slope of the chords defining this diameter.

Solution. By theorem XI the diameter has the equation $y = 2p/m$ or $y = 3/m$. But $y = -2$ is also the equation of this diameter and hence $-2 = 3/m$ or $m = -\tfrac{3}{2}$.

EXERCISES

1. In the conic $4x^2 + y^2 = 4$ a diameter bisects the chords of slope 1. Find its equation. *Ans.* $4x + y = 0$.

2. In the conic $\dfrac{x^2}{16} + \dfrac{y^2}{9} = 1$ the slope of a diameter is 3. Find the angle between this diameter and its conjugate. (*Hint:* Find the slope of the conjugate diameter and apply the formula for the tangent of the angle between two lines.) *Ans.* $\tan\theta = -\tfrac{51}{7}$.

3. For the hyperbola $-3x^2 + 2y^2 = 6$ the family of chords $4x - y = k$ defines a diameter. Find its equation. *Ans.* $3x - 8y = 0$.

4. Find the equation of the diameter of $x^2 = 12y$ defined by the chords perpendicular to $2x - y + 5 = 0$. *Ans.* $x = -3$.

CHAPTER XIV

POLAR COORDINATES

76. Definitions. Consider a horizontal line called the *polar axis* and a point O on it called the *pole,* or origin. The polar axis may be rotated about the pole through an angle θ so as to make it pass through any point P in the plane. Call $r = OP$ the distance, or *radius vector.* Then r and θ are, by definition, *polar coordinates* of the point P and we write $P(r, \theta)$. Angular measurements may be made in radians, degrees, or other units. The line through the pole perpendicular to the polar axis we call the *co-polar axis.*

We make the following conventions regarding signs. The polar axis $P'OP$ is positively directed to the right. Thus for the angle $\theta = 0$ the radius vector of a point to the right of the pole is positive. After rotation of $P'OP$ through any angle θ, r is positive in the direction OP and negative in the direction OP'. Counterclockwise rotations give positive angles, clockwise rotations negative angles.

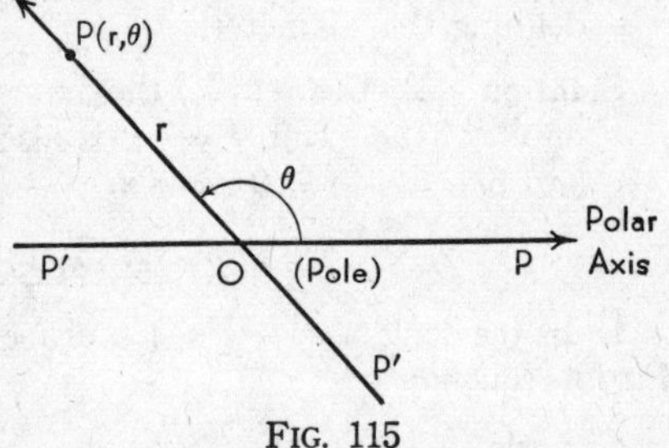

FIG. 115

The one-to-one correspondence between number pairs and points in the plane is not reciprocal as in the case of rectangular coordinates. For, corresponding to a given point, there is an infinity of coordinates: $(r, \theta + 2\,k\pi)$ and $(-r, \theta + (2\,k + 1)\pi)$ represent the same point for all integral values of k. There is no unique angle θ for the pole, where $r = 0$.

77. Relation between Polar and Rectangular Coordinates. By superposition of the two systems we see that

(1) $\begin{cases} x = r\cos\theta, \\ y = r\sin\theta; \end{cases}$

(2) $\begin{cases} r = \sqrt{x^2 + y^2}, \\ \theta = \tan^{-1}\dfrac{y}{x}. \end{cases}$

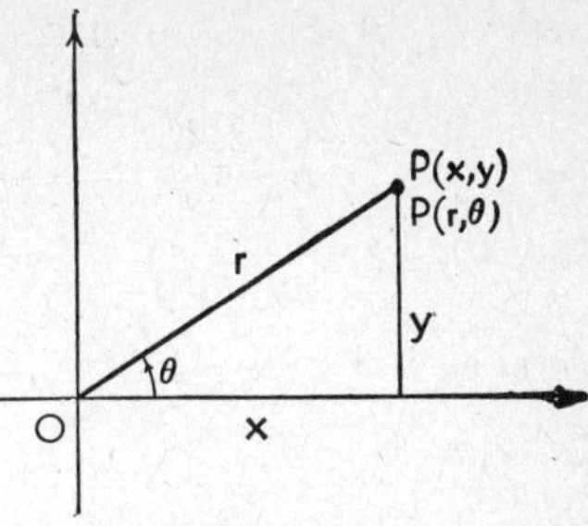

FIG. 116

The angle θ is chosen as the least (positive) angle corresponding to the quadrant in which the point lies.

(3) $$\sin\theta = \frac{y}{\sqrt{x^2 + y^2}}, \quad \cos\theta = \frac{x}{\sqrt{x^2 + y^2}}.$$

By means of these relations a given equation can be transformed from one system of coordinates to another.

Illustration 1. Transform the rectangular equation $x^2 + y^2 = a^2$ to polar coordinates.

Solution. From (1) we get

$$\begin{aligned} (r\cos\theta)^2 + (r\sin\theta)^2 &= a^2, \\ r^2(\cos^2\theta + \sin^2\theta) &= a^2, \\ r &= \pm a. \end{aligned}$$

The locus of $r = a$ is a circle of radius a with center at the pole; $r = -a$ plots the same locus (but beginning at a different point).

Illustration 2. Find the rectangular equation of the curve whose polar equation is $r = \dfrac{4}{1 + 2\cos\theta}$.

Solution. From (2) and (3) we get

$$\sqrt{x^2 + y^2} = \frac{4}{1 + 2\dfrac{x}{\sqrt{x^2 + y^2}}},$$

$$\begin{aligned} \sqrt{x^2 + y^2} + 2x &= 4, \\ x^2 + y^2 &= (4 - 2x)^2, \\ 3x^2 - y^2 - 16x + 16 &= 0 \text{ (hyperbola)}. \end{aligned}$$

78. Distance between Two Points in Polar Coordinates. The law of cosines [see p. 8, § 4, (9), (b)] applied to the triangle in Fig. 117 yields immediately the square of the distance between two points $P_1(r_1, \theta_1)$ and $P_2(r_2, \theta_2)$.

$$d^2 = r_1^2 + r_2^2 - 2\, r_1 r_2 \cos(\theta_2 - \theta_1).$$

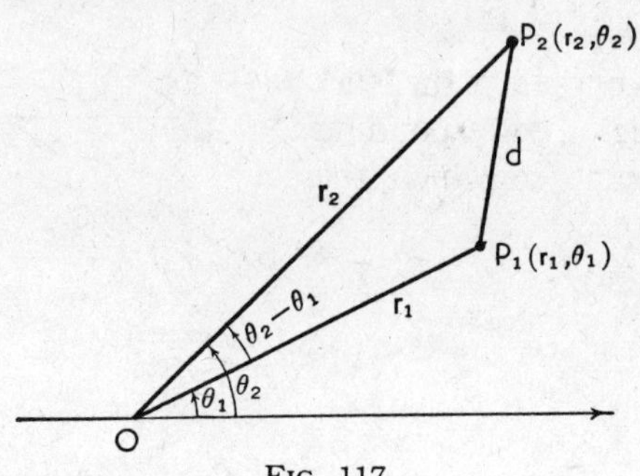

FIG. 117

This is independent of the order in which the points are chosen since $\cos(\theta_2 - \theta_1) = \cos(\theta_1 - \theta_2)$.

79. Polar Equation of a Straight Line. The equation $Ax + By + C = 0$ transforms into

$$Ar \cos \theta + Br \sin \theta + C = 0,$$

which is the general equation of a straight line in polar coordinates. Note that it is *not* linear — θ enters trigonometrically. Correspondence between rectangular and polar equations of certain special cases is established in the following table. (See § 30.)

Equation of a Straight Line

RECTANGULAR COORDINATES		POLAR COORDINATES
(1) $Ax + By + C = 0$	General	$r(A \cos \theta + B \sin \theta) + C = 0$
(2) $x \cos \phi + y \sin \phi - p = 0$	Normal	$r \cos(\theta - \phi) - p = 0$
(3) $y = mx$	Through origin	$\theta = \tan^{-1} m$
(4) $x = k$	Perpendicular to polar axis	$r = k \sec \theta$
(5) $y = k$	Parallel to polar axis	$r = k \csc \theta$

It is simple to derive any of these polar equations directly from a figure. For example the polar normal form can be read off instantly from Fig. 118.

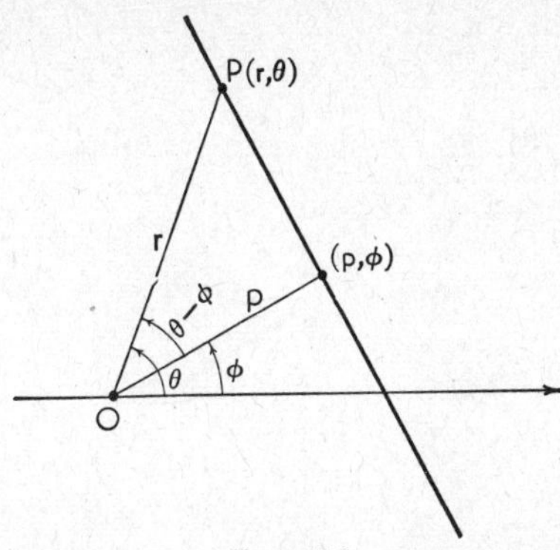

FIG. 118

80. Polar Equation of a Circle. The distance formula in § 78 enables us to write down the equation of the circle with center at (r_1, θ_1) and radius a. Directly from Fig. 119 we have

$$a^2 = r^2 + r_1^2 - 2\,rr_1 \cos(\theta - \theta_1).$$

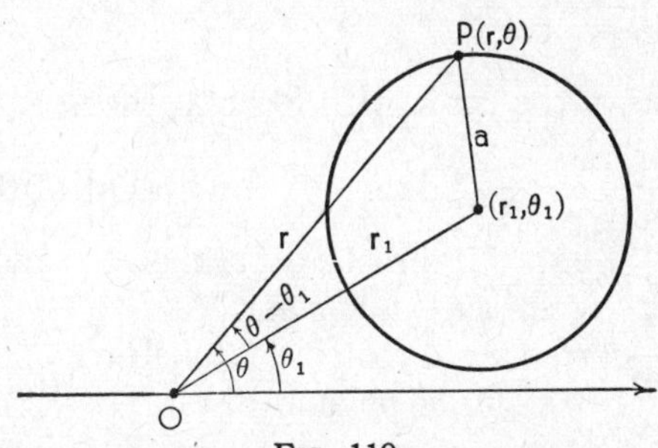

FIG. 119

The following special cases are of interest.

$r = a,$	Center at the pole, radius a;
$r = 2\,a \cos\theta,$	Center at $(a, 0°)$, radius a;
$r = 2\,a \sin\theta,$	Center at $(a, 90°)$, radius a.

81. Polar Equation of a Conic. Let the focus be at the pole and the directrix be perpendicular to the polar axis at a distance p to the left of the pole. Then (Fig. 120)

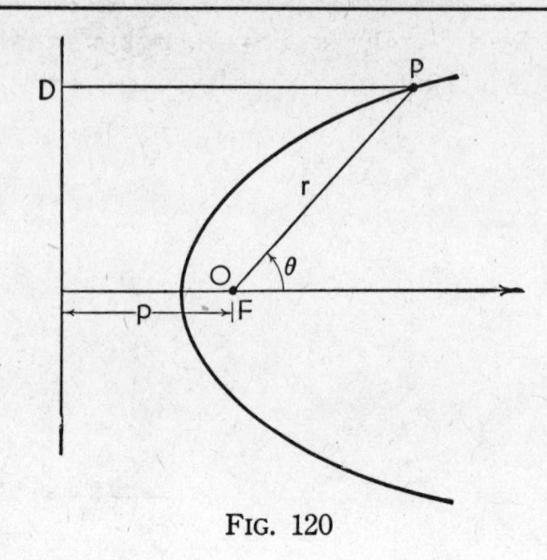

FIG. 120

$$\frac{PF}{PD} = e,$$

$$\frac{r}{p + r\cos\theta} = e.$$

Solving this equation for r we get the equation of the conic

(1) $$r = \frac{ep}{1 - e\cos\theta}.$$

Other forms of the conic with focus at the pole are

(2) $r = \dfrac{ep}{1 + e\cos\theta}$, directrix to the right of the pole;

(3) $r = \dfrac{ep}{1 - e\sin\theta}$, directrix parallel to and below the polar axis;

(4) $r = \dfrac{ep}{1 + e\sin\theta}$, directrix parallel to and above the polar axis.

Illustration. Identify and sketch $r = \dfrac{4}{1 - \frac{1}{3}\cos\theta}$.

Solution. Here $e = \frac{1}{3}$, $p = 12$. The curve is an ellipse. The ends of the major axis are on the polar axis at the points (6, 0°) and (3, 180°). The center is at ($\frac{3}{2}$, 0°). The curve can be plotted point by point in polar coordinates; or use may be made of the information gained in the study of conics in rectangular coordinates. We know that $a^2e^2 = a^2 - b^2$, whence $\frac{9}{4} = \frac{81}{4} - b^2$ or $b^2 = 18$. The rectangular equation is

$$\frac{(x - \frac{3}{2})^2}{\frac{81}{4}} + \frac{y^2}{18} = 1.$$

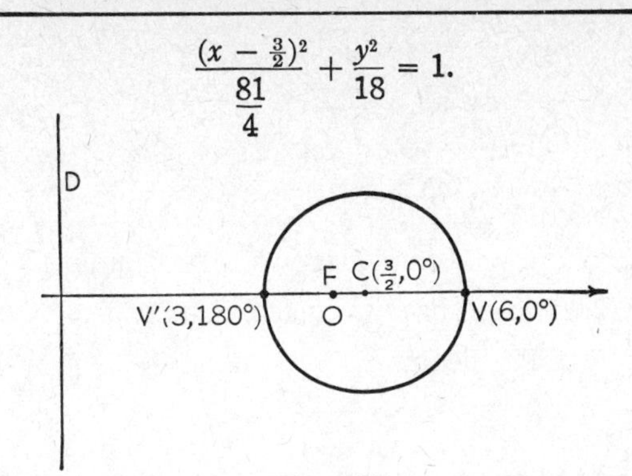

FIG. 121

EXERCISES

1. Find the distance between (1, 0°) and (1, 120°). *Ans.* $\sqrt{3}$.

2. Write the equation of the line perpendicular to the polar axis at a distance of 3 units to the left of the pole. *Ans.* $r = -3 \sec \theta$.

3. (a) Find the polar equation of the circle with center at (2, 30°) and radius 2. (b) Transform this to rectangular coordinates.

Ans. (a) $r^2 - 4r\cos(\theta - 30°) = 0$; (b) $(x - \sqrt{3})^2 + (y - 1)^2 = 4$.

4. (a) Identify and sketch $r = \dfrac{1}{1 - \sin\theta}$. (b) Transform to rectangular coordinates.

Ans. (a) Parabola; (b) $x^2 = 2(y + \frac{1}{2})$.

82. Several Equations of One Graph. A curve in polar coordinates may have more than one equation. A given point (r, θ) may have either of the two general coordinate representations

$$(1) \qquad (r, \theta + 2k\pi),$$
$$(2) \qquad (-r, \theta + (2k + 1)\pi),$$

for any integer k. Hence a given curve $r = f(\theta)$ may have either of the two equational forms

$$(A) \qquad r = f(\theta + 2k\pi),$$
$$(B) \qquad -r = f(\theta + (2k + 1)\pi).$$

Equation (A) reduces to $r = f(\theta)$ for $k = 0$ but may lead to an entirely different equation of the same curve for another value of k. Similarly (B) may yield still other equations of the curve.

Illustration 1. The polar equation of a straight line is unique.

Solution. We use the general form

(1) $$r(A \cos \theta + B \sin \theta) + C = 0.$$

Since $\cos(\theta + 2k\pi) = \cos\theta$ and $\sin(\theta + 2k\pi) = \sin\theta$, equation (A) reduces to (1). Again $\cos(\theta + (2k+1)\pi) = -\cos\theta$ and $\sin(\theta + (2k+1)\pi) = -\sin\theta$; thus equation (B) reduces to (1). There is, therefore, only one equation of a straight line.

Illustration 2. A conic has two equations in polar coordinates.

Solution. Consider the form

(1) $$r = \frac{ep}{1 - e\cos\theta}.$$

Equation (A) leads to nothing new. But equation (B) becomes

(2) $$-r = \frac{ep}{1 + e\cos\theta},$$

which is quite distinct from (1). Indeed, the coordinates of a point which satisfy (1) will not satisfy (2). The curves are the same but they are traced 180° out of phase.

Illustration 3. Find all equations of the curve $r = \cos\frac{\theta}{2}$.

Solution. Using form (A):

(A) $$r = \cos\left(\frac{\theta + 2k\pi}{2}\right).$$

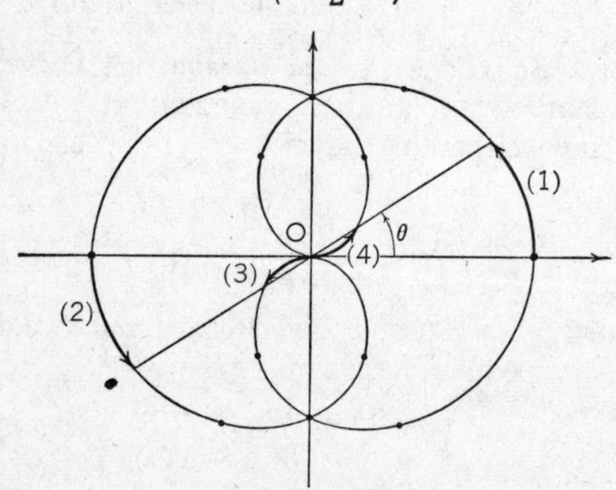

FIG. 122

This reduces to

(1) $$r = \cos\frac{\theta}{2}, \quad \text{for } k = 0;$$

(2) $$r = -\cos\frac{\theta}{2}, \text{ for } k = 1.$$

Using form (B):

$$\text{(B)} \qquad -r = \cos\left(\frac{\theta + (2k+1)\pi}{2}\right).$$

This reduces to

$$\text{(3)} \qquad r = -\sin\frac{\theta}{2}, \quad \text{for } k = 0;$$

$$\text{(4)} \qquad r = \sin\frac{\theta}{2}, \quad \text{for } k = 1.$$

Hence there are four distinct equations of this curve, which may be traced by beginning ($\theta = 0$) on any one of the four branches as shown in Fig. 122. The starting branch is indicated by the number corresponding to the associated equation. (Also see Illustration 2, § 83.)

It is useful to know the various equations of a curve when plotting; it is necessary to consider forms (A) and (B) of the equations when determining the points of intersection of loci in polar coordinates.

83. Curve Tracing in Polar Coordinates. Essentially the same procedures are followed in plotting a curve in polar as in rectangular coordinates. Intercepts, extent, symmetry, and asymptotes are investigated.

I. *Intercepts.* For the intercepts on the polar axis, set $\theta = k\pi$ and determine r. As many values of k (an integer) should be used as produce different values of r.

For the intercepts on the co-polar axis, set $\theta = \dfrac{2k+1}{2}\pi$ and determine the different values of r.

II. *Extent.* Normally θ will be thought of as the independent variable, which therefore can range from $-\infty$ to $+\infty$. The dependent variable r will plot wherever its value is real and finite. In many of the most important problems in polar coordinates r turns out to be a periodic function of θ, owing in general to the natural presence of trigonometric functions. The student should review his trigonometry at this time.

III. *Symmetry.* Let the several equations of the curve be known and let $f_i(r, \theta) = 0$, $f_j(r, \theta) = 0$ be generic notations for any of them. Then the curve will be symmetric with respect to

1. The polar axis if $f_i(r, \theta) \equiv f_j(r, -\theta)$;
2. The co-polar axis if $f_i(r, \theta) \equiv f_j(-r, -\theta)$;
3. The pole if $f_i(r, \theta) \equiv f_j(-r, \theta)$.

Converse statements also hold. The functions f_i and f_j may be the same or different.

IV. *Asymptotes.* The determination of the asymptotes is often difficult. On the other hand it is relatively simple to determine the *direction* of an asymptote: if $r = \pm\infty$ for $\theta = \theta_1$, then there is an asymptote which makes an angle θ_1 with the polar axis. Further it may happen that $\lim_{\theta \to \infty} r = k$, in which case we say that $r = k$ is an asymptote or that r approaches k asymptotically. Fortunately the question of asymptotes arises only infrequently in problems that are best suited to treatment in polar coordinates.

Illustration 1. Sketch $r = a(1 - \cos\theta)$.

Solution. Think of a as being a positive number. The intercepts are $(0, 0)$ and $(2a, \pi)$. The largest value of r is $2a$ and the curve is symmetric with respect to the polar axis since $\cos\theta = \cos(-\theta)$. Further, since $\cos\theta$ is of period 2π, the graph will be periodic of period (not exceeding) 2π. Form (B) of the equation is (see § 82) $r = -a(1 + \cos\theta)$, which reveals no other symmetry.

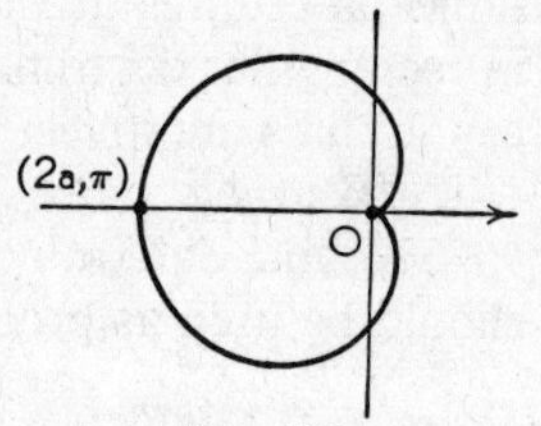

FIG. 123

Of prime importance is the reasoning exhibited in the following table.

QUADRANT	θ	$\cos\theta$	r
	0	1	0
I	$\pi/2$	0	a
II	π	-1	$2a$
III	$\frac{3}{2}\pi$	0	a
IV	2π	1	0

As θ runs through the first quadrant, $\cos\theta$ decreases from 1 to 0 and r increases from 0 to a. The other variations involved can be seen at a glance, and only a few individual points will be needed to plot the graph accurately. The figure is a cardioid.

Illustration 2. Sketch $r = \cos\frac{\theta}{2}$. (See Illustration 3, § 82.)

Solution. The first thing to note is that θ will have to run through 720° in order that $\cos\frac{\theta}{2}$ may run through a complete period.

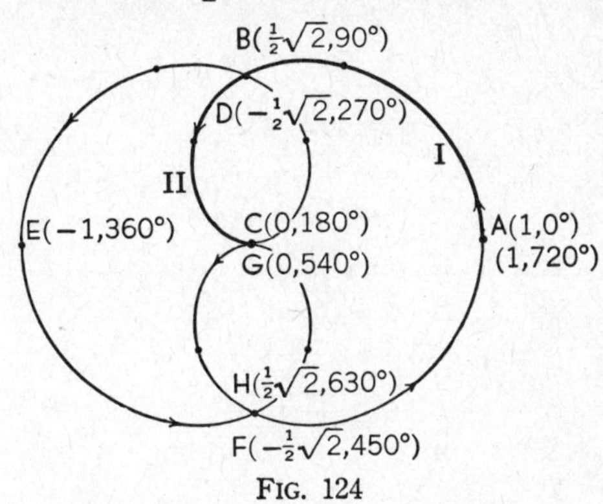

FIG. 124

The intercepts are $A(1, 0°)$, $B\left(\frac{\sqrt{2}}{2}, 90°\right)$, $E(-1, 360°)$, $F\left(-\frac{\sqrt{2}}{2}, 450°\right)$, and the pole $C(0, 180°)$. The curve has the four equations $r = \pm\cos\frac{\theta}{2}$, $r = \pm\sin\frac{\theta}{2}$. There is symmetry with respect to

1. The polar axis since $r = \cos\frac{\theta}{2} = \cos\left(-\frac{\theta}{2}\right)$;
2. The co-polar axis since $r = \sin\frac{\theta}{2}$ and $-r = \sin\left(-\frac{\theta}{2}\right)$ are the same;
3. The pole since $r = \cos\frac{\theta}{2}$ and $-r = \cos\frac{\theta}{2}$ are both equations of the curve.

The following table and a few points will enable us to sketch the graph.

QUADRANT FOR θ	θ	$\frac{\theta}{2}$	r
	0	0	1
I	90°	45°	$\frac{\sqrt{2}}{2}$
II	180°	90°	0

We apply our knowledge of symmetry for the rest of the curve and remember that r is plotted against θ, not $\frac{\theta}{2}$.

Illustration 3. Sketch $r = 1 - \frac{1}{1+\theta}$.

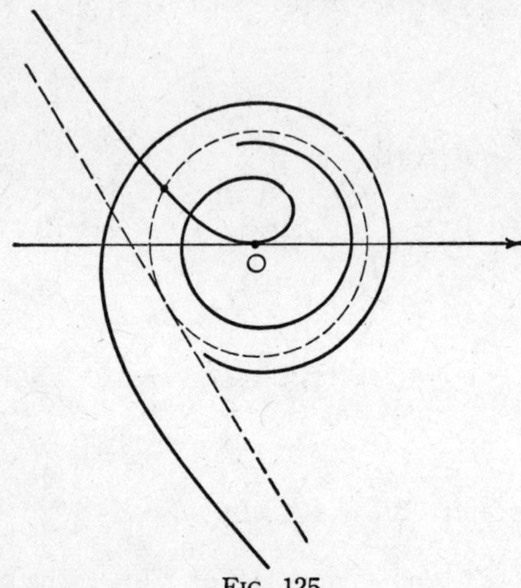

FIG. 125

Solution. Radian measure must be used.

When $\theta = 0$, $r = 0$. When $\theta = k\pi$, $r = 1 - \frac{1}{1 + k\pi}$; there are, therefore, infinitely many intercepts on the polar axis. Similarly for the co-polar axis. There is an infinity of equations of each type (A) and (B). There is no symmetry.

For $\theta > 0$ it is easy to see that r increases with θ and that $\lim_{\theta \to +\infty} r = 1$. The graph approaches the circle $r = 1$ asymptotically from the inside.

For $\theta < 0$ the curve is discontinuous at $\theta = -1$. The curve is asymptotic to a line which makes an angle of -1 (radian) with the polar axis. This asymptote, found by the aid of the calculus, is dotted in. Since $\lim_{\theta \to -\infty} r = 1$, the curve is asymptotic to the circle $r = 1$ from the outside.

The following typical graphs and their equations will serve as excellent exercises in curve tracing. The student should analyze each equation, using the graph shown as a check against his own work.

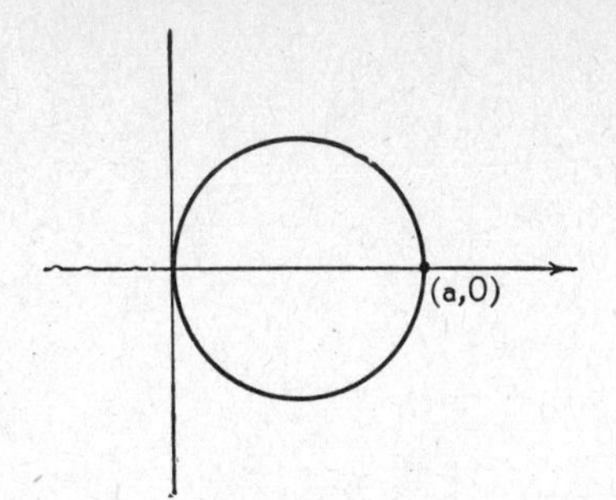

FIG. 126. Circle: $r = a \cos \theta$.

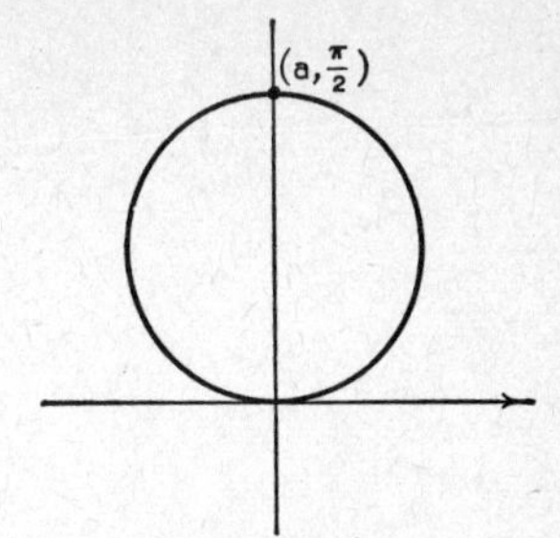

FIG. 127. Circle: $r = a \sin \theta$.

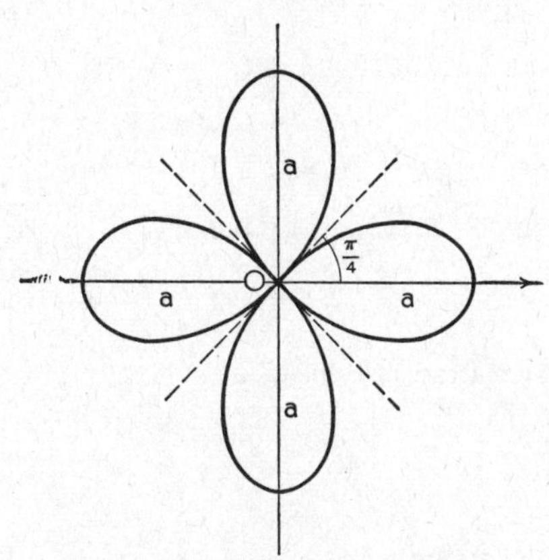

FIG. 128. Four-Leaved Rose: $r = a \cos 2\,\theta$.

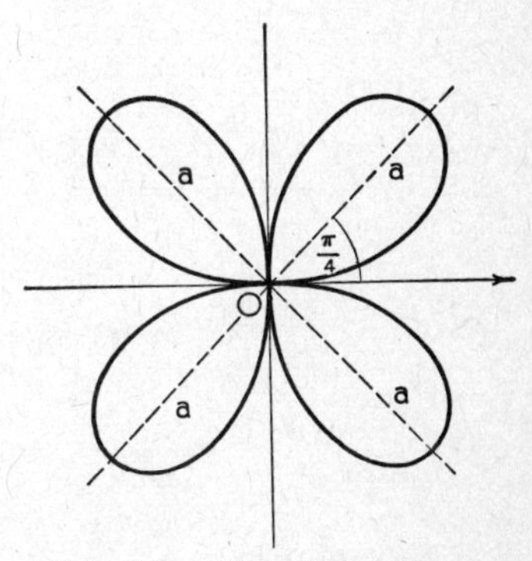

FIG. 129. Four-Leaved Rose: $r = a \sin 2\,\theta$.

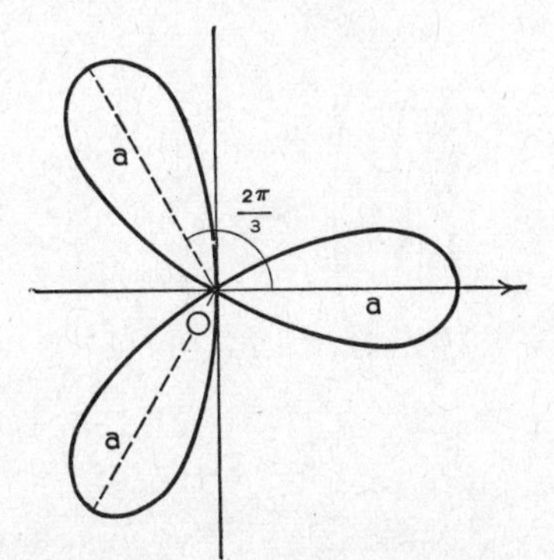

FIG. 130. Three-Leaved Rose: $r = a \cos 3\,\theta$.

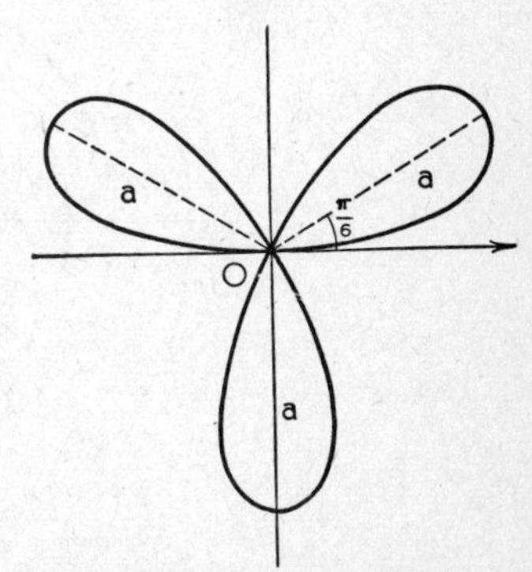

FIG. 131. Three-Leaved Rose: $r = a \sin 3\,\theta$.

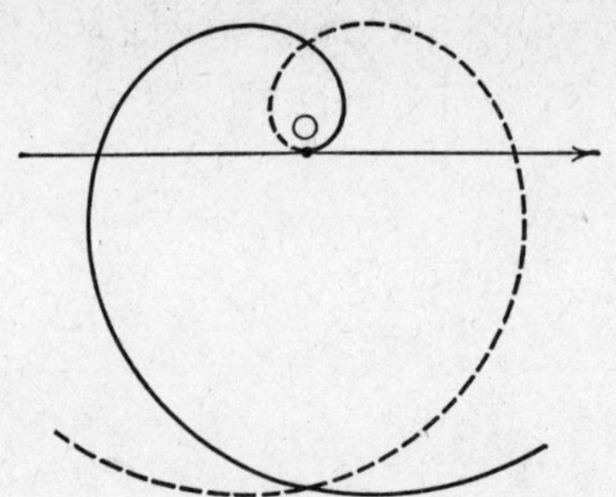

FIG. 132. The Spiral of Archimedes:
$r = a\,\theta.$

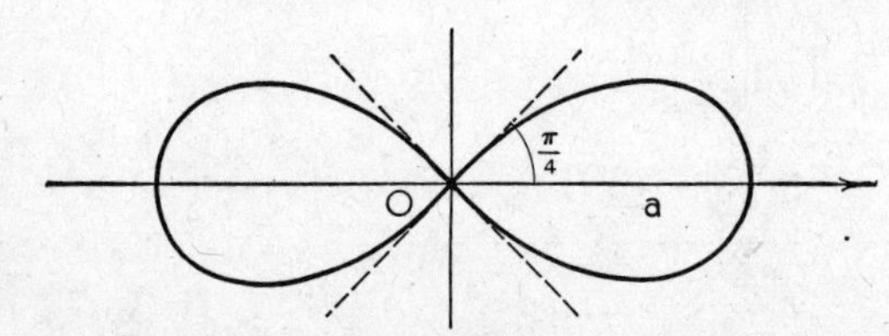

FIG. 133. The Lemniscate of Bernoulli:
$r^2 = a^2 \cos 2\,\theta.$

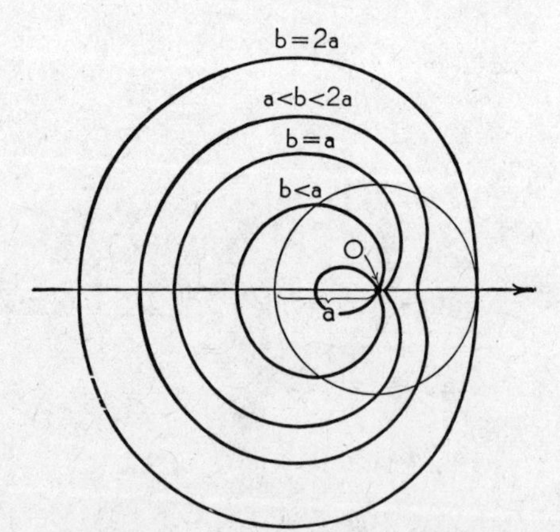

FIG. 134. The Limaçon of Pascal:
$r = b - a \cos \theta.$
(Cardioid if $b = a$. See FIG. 123.)

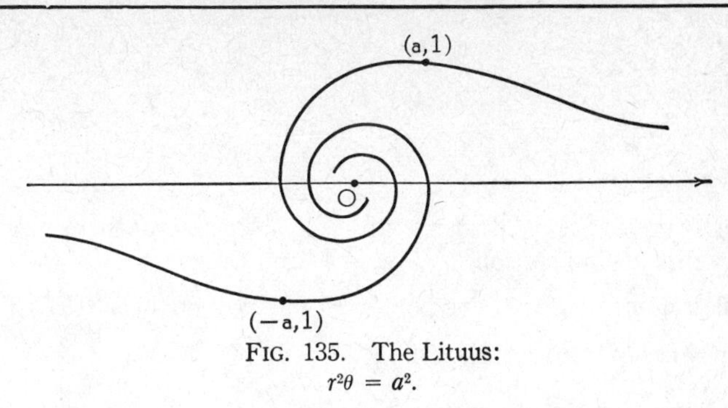

FIG. 135. The Lituus:
$r^2\theta = a^2$.

EXERCISES

Trace the following curves.

1. The Hyperbolic Spiral: $r = a/\theta$.
2. The Cruciform: $r = 2\,a \csc 2\,\theta$.
3. The Conchoid: $r = a \sec \theta \pm b$.
4. The Strophoid: $r = a(\sec \theta \pm \tan \theta)$.
5. The Trisectrix: $r = a(4 \cos \theta - \sec \theta)$.

84. Intersection of Curves in Polar Coordinates. By a point of intersection of two curves we mean a *common geometric point* regardless of the way in which the coordinates are assigned. With the exception of the pole, coordinates of a point on a curve will satisfy *some* equation of the curve. To find the points common to two curves $r = f(\theta)$ and $r = g(\theta)$ we solve simultaneously each equation of form (A) of one of them with every equation of form (B) of the other. That is, we find the solutions common to

(A) $$r = f(\theta + 2\,k\pi),$$
(B) $$-r = g(\theta + (2\,k' + 1)\pi),$$

for all integers k and k' that yield different equations.

This process may or may not lead to duplications in the answers. Often the work may be simplified or shortened by the use of symmetry, when present, or of other information. The pole is handled independently.

Illustration 1. Find the points of intersection of $r = \cos \theta$ and $r = \sin \theta$. (Figs. 126 and 127.)

Solution. Each curve has only the one equation.

$$\cos\theta = \sin\theta,$$
$$\theta = \frac{\pi}{4},$$
$$r = \frac{\sqrt{2}}{2}.$$

Each curve passes through the pole. Therefore there are two points of intersection, $(\sqrt{2}/2, \pi/4)$ and the pole.

Illustration 2. Find the points of intersection of $r = \cos\theta$ and $r = \cos\frac{\theta}{2}$. (Figs. 126 and 124.)

Solution. The first curve has only the one equation. The second curve has two equations of form (A), $r = \pm\cos\frac{\theta}{2}$, and two of form (B), $r = \pm\sin\frac{\theta}{2}$. (See Illustration 3, § 82.) We may use either form in solving with $r = \cos\theta$.

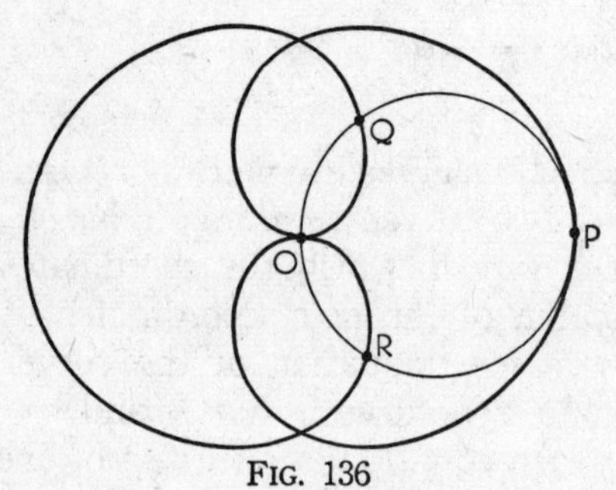

FIG. 136

Using form (A) we get:

$$r = \cos\theta, \qquad r = \cos\theta,$$
$$r = \cos\frac{\theta}{2}, \qquad r = -\cos\frac{\theta}{2},$$
$$P(1, 0°),\ Q(-\tfrac{1}{2}, 240°); \qquad P'(1, 360°),\ R(-\tfrac{1}{2}, 120°).$$

Using form (B) we get:

$$r = \cos\theta, \qquad r = \cos\theta,$$
$$r = \sin\frac{\theta}{2}, \qquad r = -\sin\frac{\theta}{2},$$
$$Q'(\tfrac{1}{2}, 60°),\ R'(\tfrac{1}{2}, 300°),\ P''(-1, 180°); \qquad P''(-1, 180°).$$

There are four points of intersection: the pole and the points P, Q, R (with or without primes) regardless of what coordinates are used.

Note that either method is adequate but that there is duplication in each.

EXERCISES

Find the points of intersection of

1. $r = \sin\theta$, $r = 1 - \sin\theta$. *Ans.* $(\frac{1}{2}, 30°)$, $(\frac{1}{2}, 120°)$, pole.
2. $r = 2\sin\theta$, $r = 2\cos 2\theta$. *Ans.* $(1, \pi/6)$, $(1, 5\pi/6)$, $(-2, 3\pi/2)$, pole.
3. $r\theta = 1$. $\theta = 1$.

   Ans. $\left(\frac{1}{1 + 2k\pi}, 1\right)$, $\left(\frac{-1}{1 + (2k+1)\pi}, 1\right)$, for all integral values of k.

85. Loci in Polar Coordinates. The basic principles involved in loci problems are the same regardless of the coordinate system used. (See § 21.) We seek the locus of a point which moves in accordance with certain prescribed geometric conditions. The equation is determined by expressing these conditions analytically in terms of the coordinates (r, θ) of the moving point. By careful selection of the polar axis and pole the intermediate work can often be reduced and the final equation simplified.

Illustration 1. Find the locus of P which moves so that its radius vector is proportional to the square of its vectorial angle.

Solution. The answer is immediate: the locus (a spiral) is given by

$$r = k\theta^2$$

where k is the factor of proportionality.

Illustration 2. Find the locus of the midpoints of chords of a circle of radius a drawn from a fixed point O on the circle.

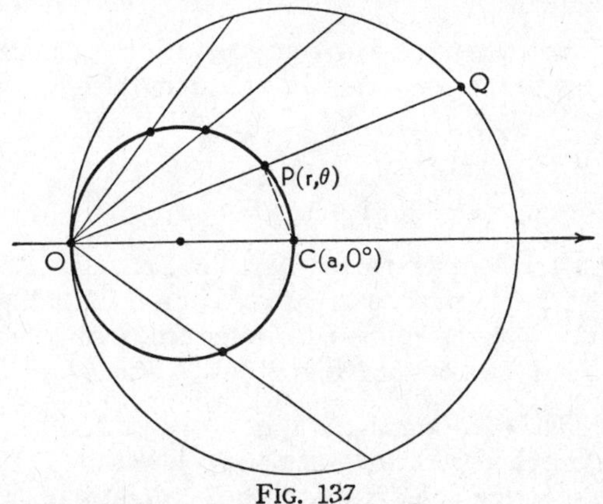

FIG. 137

Solution. Let the polar axis coincide with a diameter through O and choose O as the pole. The midpoint of a general chord OQ we label $P(r, \theta)$ and draw PC where C is the center of the circle. The coordinates of C are $(a, 0°)$ and OPC is a right triangle. Thus

$$r = a \cos \theta$$

is the equation of the locus, a circle with center at $\left(\frac{a}{2}, 0°\right)$ and radius $a/2$.

Illustration 3. Consider a circle of radius a and a diameter OA. A line OQ is drawn intersecting the circle at Q. Let P be a point on the line OQ such that $OP = OQ + 2a$. Find the locus of P.

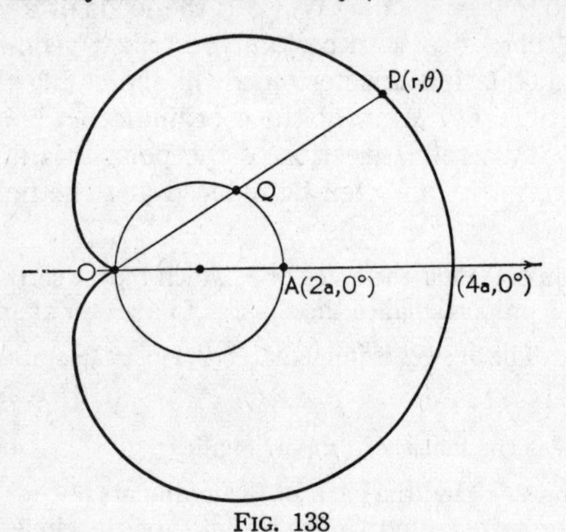

FIG. 138

Solution. Let the polar axis coincide with the diameter OA and the pole with the point O. Directly from the figure we have

$$OP = OQ + 2a,$$
$$r = 2a \cos \theta + 2a,$$
$$r = 2a(1 + \cos \theta), \text{ a cardioid.}$$

Illustration 4. Consider a circle of radius $2a$, a diameter OCA, and a line L perpendicular to OA erected at a point halfway between O and C. A variable line OBD intersects L at B and the circle at D. Find the locus of P on OBD such that $OP = BD$.

Solution. The equation of L will be $r = a \sec \theta$ if the polar axis is taken coincident with OCA and the pole with O. The equation of the circle is $r = 4a \cos \theta$. The coordinates of P, B, and D

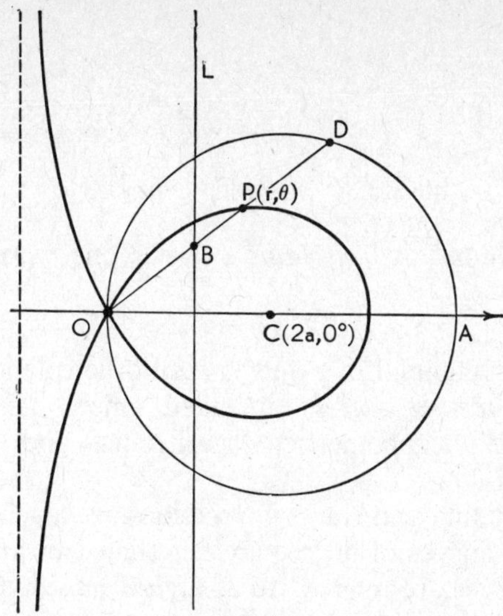

FIG. 139

are $P(r, \theta)$, $B(a \sec \theta, \theta)$, and $D(4\,a \cos \theta, \theta)$. The condition $OP = BD$ becomes

$$r = 4\,a \cos \theta - a \sec \theta,$$
$$r = a(4 \cos \theta - \sec \theta),$$

which is the equation of the trisectrix. (See Exercise 5, § 83.)

EXERCISES

1. Through the focus F of a conic, focal radii FP' are drawn, P' being on the conic. The point P is taken on FP' so that FP is proportional to FP'. Show that the locus of P is a conic with the same focus and eccentricity.

2. A variable line OA through the pole cuts the circle $r = a \cos \theta$ at B and the straight line $r = a \sec \theta$ at C. Find the locus of P if P is on OA and if $OP = BC$. *Ans.* The cissoid, $r = a(\sec \theta - \cos \theta)$.

3. A variable line L through the pole cuts the fixed line $r = a \sec \theta$ at A. Along L and from A in either direction the distance $AP = b$ is laid off. Find the locus of P.

Ans. The conchoid, $r = a \sec \theta \pm b$. (See Exercise 3, § 83.)

4. A variable line L through the pole cuts the fixed line $r = a \sec \theta$ at A. Let the projection of A on the polar axis be B. Along L and from A in either direction the distance $AP = AB$ is laid off. Find the locus of P.

Ans. The strophoid, $r = a(\sec \theta \pm \tan \theta)$. (See Exercise 4, § 83.)

CHAPTER XV

HIGHER PLANE CURVES

86. Definitions. An *algebraic curve* is one whose equation is of the form

$$f(x, y) = 0$$

where f is a polynomial in x and y All other plane curves are called *transcendental curves;* included among them are the graphs of the trigonometric, logarithmic, and exponential functions.

The straight line and the conic sections are algebraic curves. The algebraic curves of degree greater than two and the transcendental curves are referred to as *higher plane curves.*

87. Algebraic Curves. We have already treated a number of higher plane curves in our study of polar coordinates. The ones studied were especially amenable to such treatment. We now consider several in rectangular coordinates.

Illustration 1. A tangent is drawn at one end of a diameter of a circle of radius a. A variable line through the other end of the diameter cuts the circle at A and the tangent at B. A line through A parallel to the tangent and a line through B parallel to the diameter meet in the point P. Find the locus of P.

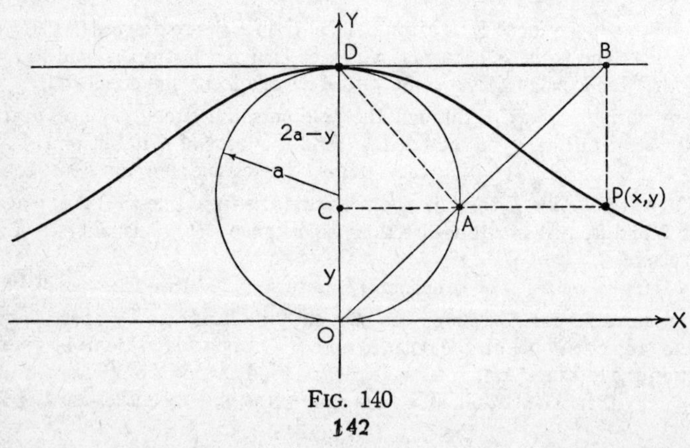

FIG. 140

Solution. We choose axes as in Fig. 140. Now

$$(1) \qquad OC : CA = OD : DB.$$

Further $CA = \sqrt{y(2a - y)}$ since it is the mean proportional between y and $2a - y$. Thus (1) becomes

$$y : \sqrt{y(2a - y)} = 2a : x.$$

This reduces to

$$x^2y = 4a^2(2a - y).$$

The curve (a cubic) is known as the Witch of Agnesi.

Illustration 2. A variable line L through the origin cuts the fixed line $x = a$ at A. Along L and from A in either direction the distance b is laid off, the end points being P and P'. Find the locus of P and P'.

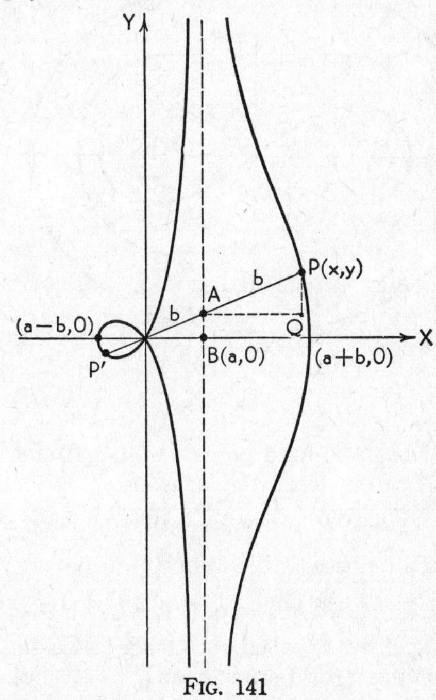

FIG. 141

Solution. In Fig. 141 we have

$$\frac{AB}{a} = \frac{y}{x},$$

$$AB = \frac{ay}{x};$$

$$PQ = y - \frac{ay}{x} = \frac{y}{x}(x - a);$$

$$(x - a)^2 + \frac{y^2}{x^2}(x - a)^2 = b^2.$$

This reduces to

$$(x^2 + y^2)(x - a)^2 = x^2b^2.$$

The curve (a quartic) is called the Conchoid of Nicomedes. (See Exercise 3, § 85.)

Illustration 3. Find the locus of a point which moves so that the product of its distances from two fixed points is a constant.

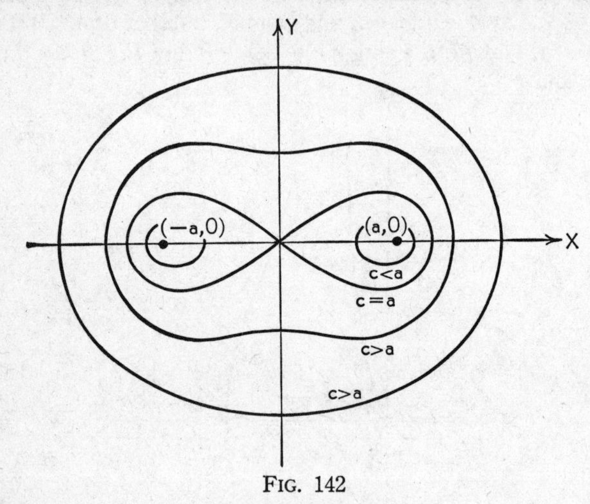

FIG. 142

Solution. Let the two fixed points be $(-a, 0)$ and $(a, 0)$ and let the constant be c^2. Then the locus is

$$\sqrt{(x + a)^2 + y^2}\sqrt{(x - a)^2 + y^2} = c^2,$$

which may be reduced to the form

$$(x^2 + y^2 + a^2)^2 - 4\,a^2x^2 = c^4.$$

The curve (a quartic) is called the Ovals of Cassini. (It reduces to the Lemniscate of Bernoulli for $c = a$. See Fig. 133 and Exercise 7 below.)

The first pedal curve of a given curve with respect to a given point is the locus of the foot of the normal from the point upon a variable tangent to the curve. Pedal curves of the conics are interesting higher plane curves.

Illustration 4. The first pedal curve of a parabola with respect to the point of intersection of the axis and the directrix is a strophoid.

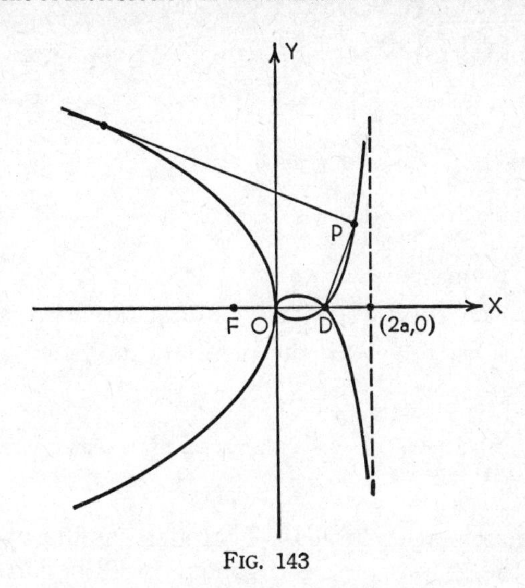

FIG. 143

Solution. Let the parabola have the equation

$$y^2 = -4\,ax.$$

The equation of the directrix is $x = a$. A tangent of slope m has the equation (see § 44)

$$y = mx - \frac{a}{m} \tag{1}$$

and the normal to this through $(a, 0)$ is

$$y = -\frac{1}{m}(x - a). \tag{2}$$

We seek the locus of the point of intersection of (1) and (2). If we eliminate m between the two equations we shall have the equation of the locus. Solving (2) for m and substituting in (1) we get

$$y = -\frac{(x - a)}{y}x + \frac{ay}{(x - a)},$$

which reduces to

$$x^3 + x(a^2 + y^2) = 2\,a(y^2 + x^2).$$

EXERCISES

Transform the following polar equations to rectangular coordinates.

1. The Cruciform: $r = 2\,a \csc 2\,\theta$. *Ans.* $x^2y^2 = a^2(x^2 + y^2)$.

2. The Trisectrix: $r = a(4 \cos \theta - \sec \theta)$. *Ans.* $y^2 = \dfrac{x^2(3\,a - x)}{a + x}$.

3. The Cissoid: $r = a(\sec \theta - \cos \theta)$. *Ans.* $y^2 = \dfrac{x^3}{a - x}$.

4. The Cardioid: $r = a(1 + \cos \theta)$. *Ans.* $(x^2 + y^2 - ax)^2 = a^2(x^2 + y^2)$.

Prove that the first pedal curve of

5. A circle with respect to a point on the circumference is a cardioid;

6. The parabola $y^2 = -4\,ax$ with respect to the vertex is the cissoid $y^2 = \dfrac{x^3}{a - x}$;

7. The hyperbola $x^2 - y^2 = a^2$ with respect to its center is the lemniscate $(x^2 + y^2)^2 = a^2(x^2 - y^2)$.

88. Trigonometric Curves. Thorough familiarity with the details of the graphs of the trigonometric functions is highly desirable. Problems involving them occur again and again in all branches of both pure and applied mathematics as well as in physics, chemistry, and engineering. The student should make use of Table 3, § 4, and Table I, Appendix B, and draw for himself these graphs, checking his own with those that follow.

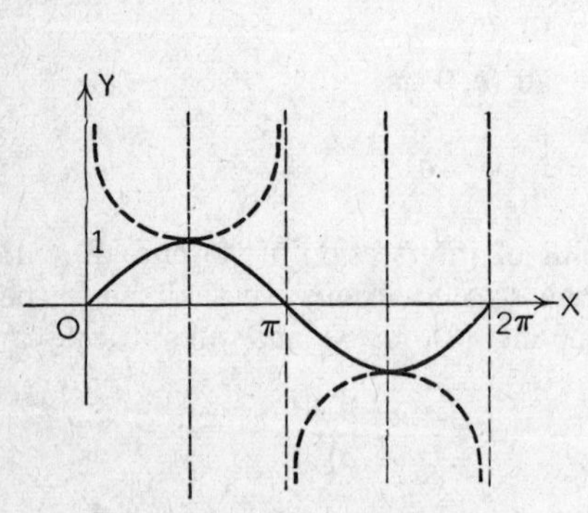

FIG. 144. $y = \sin x$ ——
FIG. 145. $y = \csc x$ ----

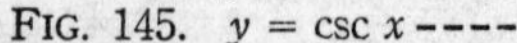

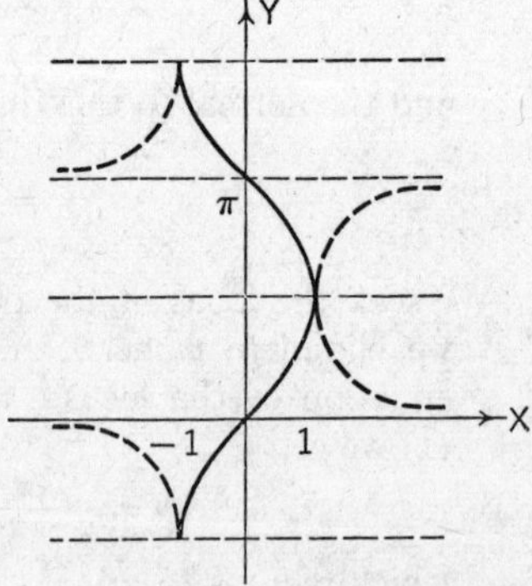

FIG. 146. $y = \sin^{-1} x$ ——
FIG. 147. $y = \csc^{-1} x$ -----

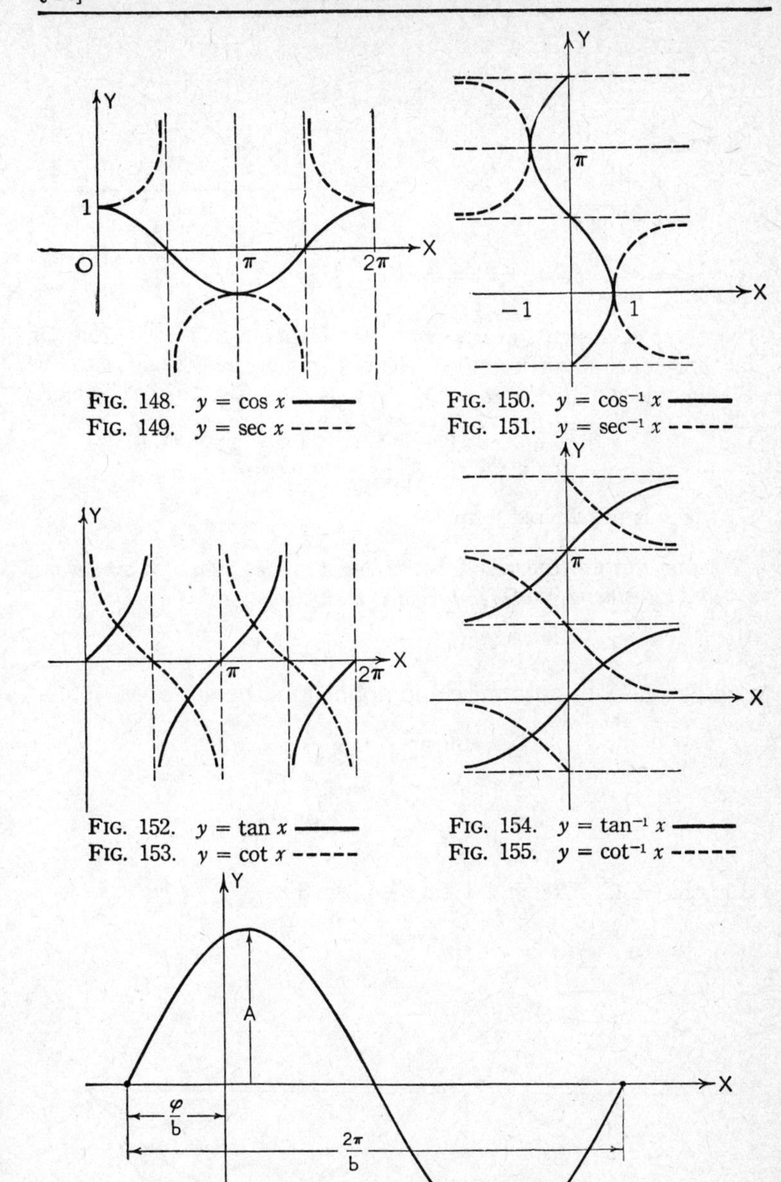

FIG. 148. $y = \cos x$ ———

FIG. 149. $y = \sec x$ - - - - -

FIG. 150. $y = \cos^{-1} x$ ———

FIG. 151. $y = \sec^{-1} x$ - - - - -

FIG. 152. $y = \tan x$ ———

FIG. 153. $y = \cot x$ - - - - -

FIG. 154. $y = \tan^{-1} x$ ———

FIG. 155. $y = \cot^{-1} x$ - - - - -

FIG. 156. Sine Wave: $y = A \sin (bx + \varphi)$.

Illustration 1. Show that $y = a \sin bx + c \cos bx$ is a sine wave of the form $y = A \sin (bx + \phi)$.

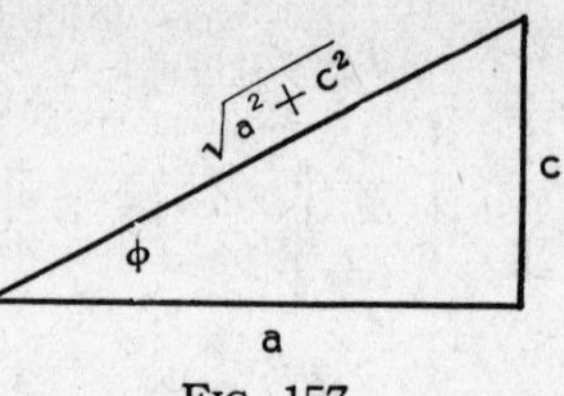

FIG. 157

Solution.

$$y = a \sin bx + c \cos bx$$
$$= \sqrt{a^2 + c^2}$$
$$\left(\frac{a}{\sqrt{a^2 + c^2}} \sin bx + \frac{c}{\sqrt{a^2 + c^2}} \cos bx\right).$$

Now the coefficients of sin bx and cos bx are such that the sum of their squares equals unity. Hence they may be thought of as the sine and cosine of some angle, say ϕ, as in Fig. 157. Therefore

$$y = \sqrt{a^2 + c^2} (\cos \phi \sin bx + \sin \phi \cos bx)$$
$$= A \sin (bx + \phi),$$

which is the desired form.

We may restate the result by saying that *the sum of a sine wave and a cosine wave of like period is a sine wave.*

$$\text{Amplitude} = A,$$
$$\text{period} = \frac{2\pi}{b},$$
$$\text{frequency} = \frac{b}{2\pi},$$
$$\text{phase angle} = \phi,$$
$$\text{phase shift} = -\frac{\phi}{b}.$$

Illustration 2. Sketch $y = \sin x + \frac{1}{3} \sin 3x$.

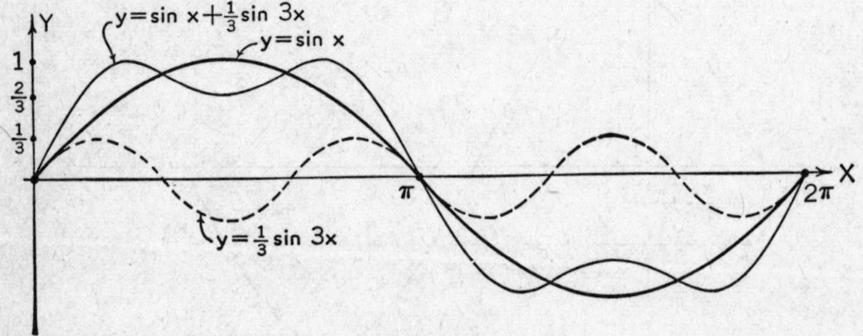

FIG. 158

Solution. We employ a standard and most useful method known as the *composition of ordinates.* First plot $y = \sin x$ and $y = \frac{1}{3} \sin 3x$ separately. We wish the sum of these two curves and it is a simple matter to take dividers and add the ordinates graphically for any given abscissa.

EXERCISES

1. Show that $y = a \sin bx + c \cos bx$ may be represented as a cosine wave of the form $y = A \cos(bx - \omega)$.

2. On the same set of axes and to the same scales plot (a) $y = \cos x/2$, (b) $y = 3 \cos 2x$, (c) $y = \cos x/2 - 3 \cos 2x$.

3. Sketch $y = \tan (x - \pi/4)$.

89. Logarithmic and Exponential Curves. From the definition of logarithm [see (4), § 2]

$$y = \log_a x \tag{1}$$

and

$$x = a^y \tag{2}$$

are identical statements of the relation that exists between x and y. Equation (1) expresses y explicitly as a function of x while (2) expresses x as a function of y. For a given base a the graphs of (1) and (2) are therefore identical.

The two principal bases in use are $a = 10$, which gives rise to *common logarithms,* and $a \doteq 2.718$, which gives rise to the natural or *Naperian logarithms.* The number 2.718 is an approximation for $\lim\limits_{n \to \infty} (1 + 1/n)^n$ and is generally designated by the letter e. Base 10 is most useful in computational work, base $e = 2.718 \cdots$ in theoretical.

In Appendix B, Tables II, III, and V are devoted to logarithms and exponentials. By means of these tables the details of plotting logarithmic and exponential functions may be facilitated.

Illustration 1. Sketch $y = \log_{10} x$.

Solution. From Table II we compute a few points: $(\frac{1}{2}, -.3010)$, $(1, 0)$, $(2, .3010)$, $(5, .6990)$. The logarithms of numbers between 0 and 1 are negative, and $x = 0$ is a vertical asymptote $(\log 0 = -\infty)$. There are no real logarithms of negative num-

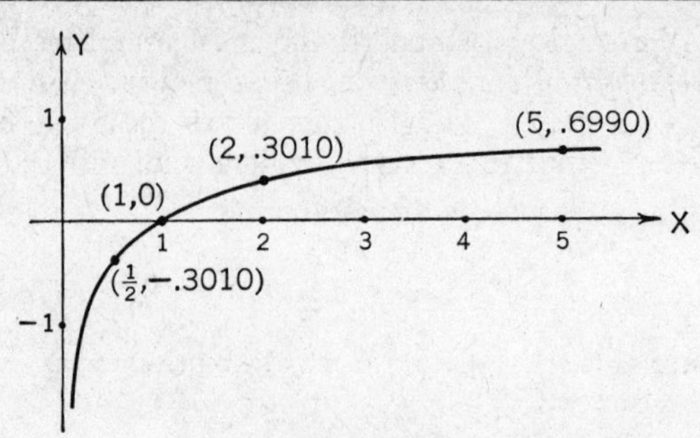

Fig. 159

bers. As $x \to +\infty$, $\log x \to +\infty$. Fig. 159 is also the graph of $x = 10^y$.

Illustration 2. Sketch (a) $y = 10^x$, (b) $y = e^x$.

Solution. (a) This graph will quite evidently be the same as that of $x = 10^y$ with the appropriate changes in the axes.

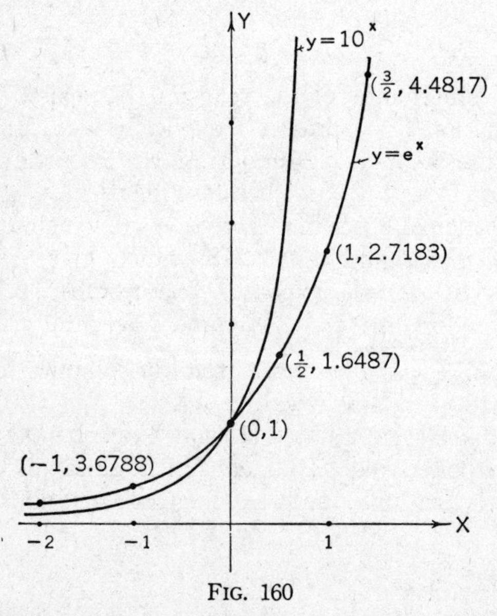

Fig. 160

(b) From Table V we compute the coordinates of the following points on $y = e^x$: $(-2, .13534)$, $(-1, .36788)$, $(0, 1)$, $(\frac{1}{2}, 1.6487)$, $(1, 2.7183)$, $(\frac{3}{2}, 4.4817)$. The graph has essentially the same character as that of $y = 10^x$.

Illustration 3. Sketch $y = e^{-x} \sin x$.

Solution. The graph may be thought of as the product of the two graphs $y = e^{-x}$ and $y = \sin x$. Thus, for any abscissa, we com-

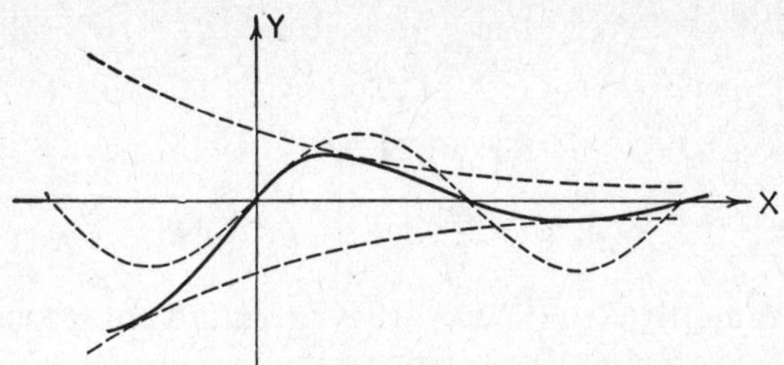

FIG. 161

pute e^{-x} and $\sin x$ and multiply these together to get the ordinate of $y = e^{-x} \sin x$. The curve is known as the *exponentially damped sine wave.*

Illustration 4. Sketch $y = e^{-x^2}$.

Solution. The curve crosses the Y-axis at (0, 1); there is no intercept on the X-axis, which is an asymptote. There is symmetry with respect to the Y-axis. By computing a few values (Table V

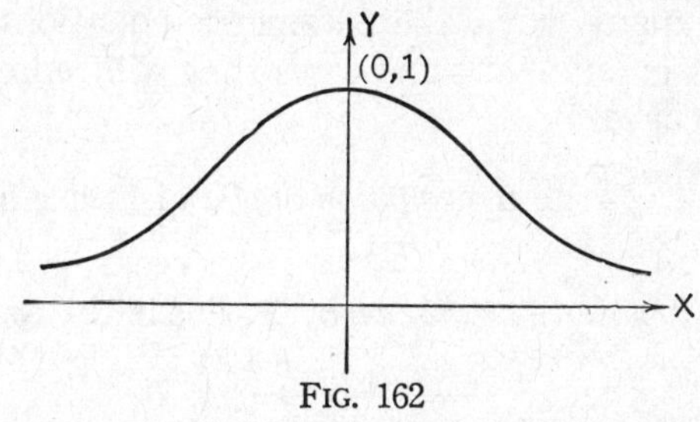

FIG. 162

may be used) an accurate graph can be constructed. The graph is known as the *probability curve* and is of great use in statistical theory.

EXERCISES

Sketch the following curves.

1. $y = \log (x + 2)$.
2. $y = \log \sqrt{x^2 - 9}$.
3. $y = - 4 e^{-2x}$.
4. $y = xe^{-x}$.
5. $y = \frac{a}{2} \left(e^{\frac{x}{a}} + e^{-\frac{x}{a}}\right)$, the catenary.

CHAPTER XVI

PARAMETRIC EQUATIONS

90. Parametric Equations. It is often advantageous to use two equations to represent a curve instead of one. The x-coordinate of a point on the curve will be given by one equation expressing x as some function of a *parameter*, say θ, or t, and the y-coordinate will be given by another equation expressing y as a function of the parameter. Such equations are called *parametric equations.* Upon eliminating the parameter between the two equations, the Cartesian equation of the curve is obtained. Many loci problems are treated most readily by means of parametric equations. Since a parameter may be chosen in many ways, the parametric equations of a given curve are not unique; in some cases they will represent only a portion of a curve.

Illustration 1. Write the equation of a straight line in parametric form.

Solution. Consider the equation of the line in the two-point form

$$\frac{y - y_1}{y_2 - y_1} = \frac{x - x_1}{x_2 - x_1}$$

and set each of these ratios equal to the parameter t. Solving for x and y we get as parametric equations of the straight line

$$x = x_1 + t(x_2 - x_1),$$
$$y = y_1 + t(y_2 - y_1).$$

Illustration 2. From Fig. 163 find the parametric equations of the ellipse $b^2x^2 + a^2y^2 = a^2b^2$ in terms of the angle θ.

Solution. It is immediately evident that

$$x = a \cos \theta,$$
$$y = b \sin \theta,$$

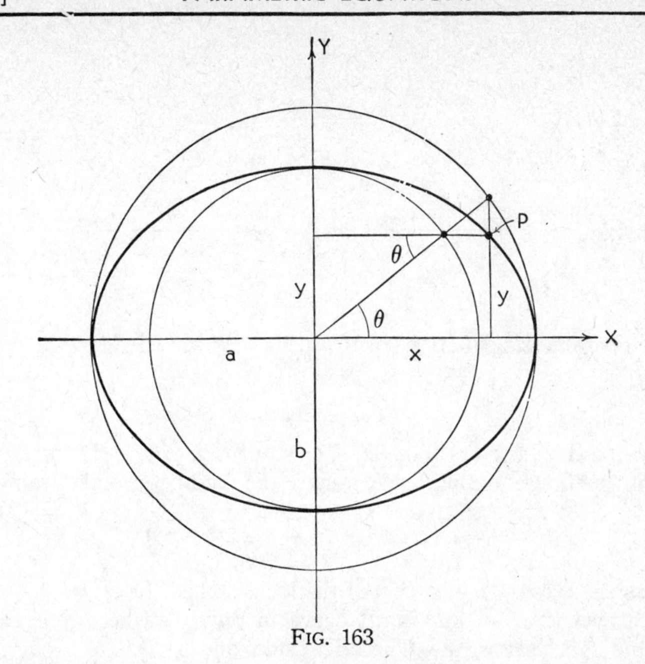

FIG. 163

which are the equations sought. If $a = b$ we have

$$x = a \cos \theta,$$
$$y = a \sin \theta$$

as parametric equations of the circle $x^2 + y^2 = a^2$.

Illustration 3. Find the curve traced by a point on the circumference of a circle of radius a as it rolls along a line.

Solution. Let the X-axis be the line and let the initial position of the tracing point P coincide with the origin. We take as parameter (Fig. 164) the angle $PCN = \theta$ (in radians) through which the radius PC turns as the wheel rolls into a typical position. We seek the locus of P. Now

$$ON = \text{arc } PN = a\theta.$$

Hence

$$\begin{aligned} x &= ON - BN \\ &= ON - PM \\ &= a\theta - a \sin \theta. \end{aligned}$$

Further

$$\begin{aligned} y &= PB = CN - CM \\ &= a - a \cos \theta. \end{aligned}$$

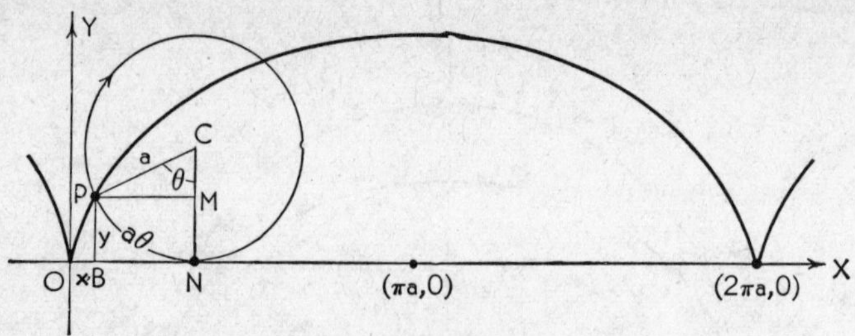

FIG. 164

The parametric equations of the locus of P are therefore

$$x = a(\theta - \sin\theta),$$
$$y = a(1 - \cos\theta).$$

Solving the second of these for θ, as a function of y, and substituting in the first equation, we eliminate the parameter and obtain

$$x = a\left(\cos^{-1}\frac{a - y}{a} - \sqrt{2\,ay - y^2}\right)$$

as the Cartesian equation of the locus, called the *cycloid*.

The cycloid is an important curve in physics, where it is called the *brachistochrone*. Turned upside down, it is the curve of steepest descent: a particle will slide down it in minimum time, the same regardless of the particular point on the curve from which the particle is released.

Illustration 4. A circle of radius b rolls on the inside of a circle of radius a. Find the locus of a point P on the circumference of the rolling circle.

Solution. Consider Fig. 165 where the parameter is θ, the angle through which the radius vector to the center C of the rolling circle has turned. First note that θ and ϕ, the angle through which the radius of the rolling wheel has turned, are connected by the relation

$$\text{arc } NA = \text{arc } NP,$$
$$a\theta = b\phi,$$

or

$$\phi = \frac{a}{b}\theta.$$

Further

$$\alpha = 90° + \theta - \phi$$
$$= 90° + \theta - \frac{a}{b}\theta$$

$$= 90° - \left(\frac{a-b}{b}\right)\theta,$$

and $$\sin\alpha = \cos\frac{a-b}{b}\theta.$$

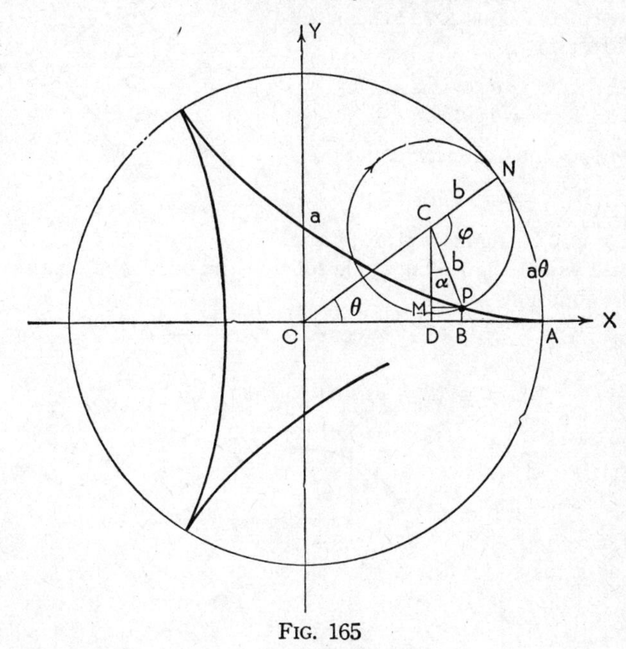

FIG. 165

The coordinates of P are x and y.

$$x = OB = OD + DB$$
$$= (ON - CN)\cos\theta + b\sin\alpha,$$

(1) $$x = (a-b)\cos\theta + b\cos\frac{a-b}{b}\theta.$$

Similarly

$$y = BP = DC - MC,$$

(2) $$y = (a-b)\sin\theta - b\sin\frac{a-b}{b}\theta.$$

The curve is called the *hypocycloid*. If a is an integral multiple of b the tracing point will return to its original position at A after the rolling circle makes one trip around the fixed circle. If a and b are commensurable the tracing point will ultimately return. But if a and b are incommensurable the tracing point will never return to its original position.

If $a = 4\,b$ the curve is the *astroid* or *hypocycloid of four cusps* (Fig. 166). It is left as an exercise for the student to show that the parametric equations in this case reduce to

$$x = a\cos^3\theta,$$
$$y = a\sin^3\theta,$$

whence the Cartesian equation is

$$x^{\frac{2}{3}} + y^{\frac{2}{3}} = a^{\frac{2}{3}}.$$

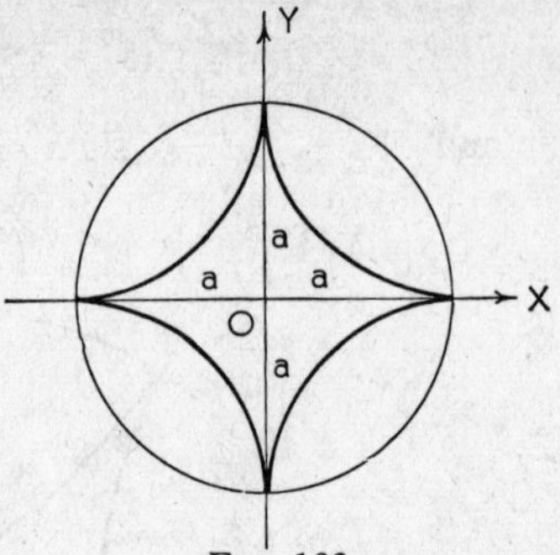

FIG. 166

By simply changing $-b$ to $+b$ the case where the rolling circle rolls on the outside is obtained and the equations are

$$x = (a + b)\cos\theta - b\cos\frac{a+b}{b}\theta,$$

$$y = (a + b)\sin\theta - b\sin\frac{a+b}{b}\theta.$$

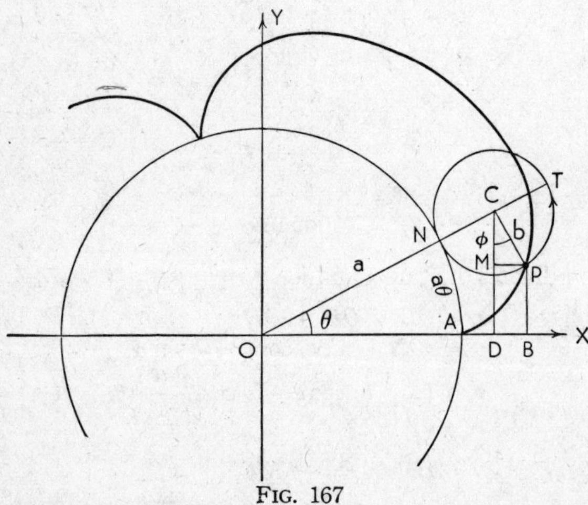

FIG. 167

The curve is the *epicycloid* (Fig. 167). All the cycloids are used in the design of gear teeth.

Illustration 5. Sketch the curve whose parametric equations are

$$x = \frac{3\,at}{(1 + t^3)},$$
$$y = \frac{3\,at^2}{(1 + t^3)}.$$

Solution. A table of values of x and y, for values of t, is constructed.

t	0	$\frac{1}{2}$	1	2	$+\infty$	$-\frac{1}{2}$	-1	-2
x	0	$4a/3$	$3a/2$	$2a/3$	0	$-12a/7$	$\pm\infty$	$6a/7$
y	0	$2a/3$	$3a/2$	$4a/3$	0	$6a/7$	$\pm\infty$	$-12a/7$

The curve, called the *Folium of Descartes*, is shown in Fig. 168 along with the range of values of the parameter t associated with each portion of the graph. The line $x + y + a = 0$ is an oblique asymptote.

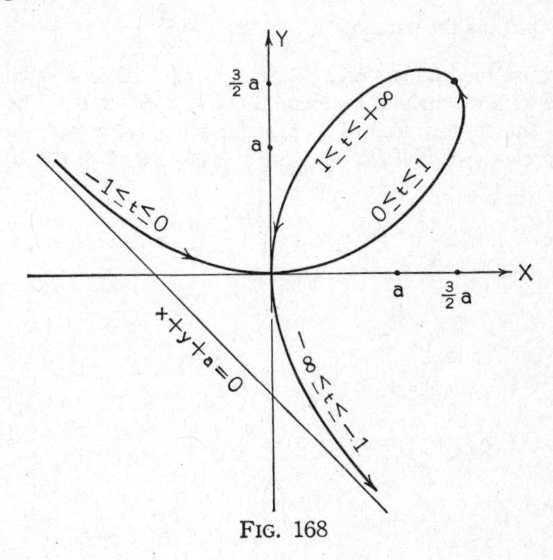

FIG. 168

By eliminating the parameter we obtain

$$x^3 + y^3 - 3\,axy = 0$$

as the Cartesian equation. The folium is an algebraic curve of the third degree. The original parametric equations can be recovered by setting $y = tx$ and solving this simultaneously with the Cartesian equation for x and y as functions of t. From this it follows that the geometric significance of this particular parameter is the *slope of the line joining the origin and a point on the curve.* Of course if y is chosen as some other function of x and t the parametric equations will be different.

EXERCISES

1. In Fig. 140 take angle $DOB = \theta$ as parameter and show that the parametric equations of the witch are $x = 2a \tan \theta$, $y = 2a \cos^2 \theta$.

2. If a string, wound around a circle, is unwound, a point on the string will trace a curve called the *involute* of the circle. Show that the parametric equations of the involute of the circle whose polar equation is $r = a$ are $x = a \cos \theta + a\theta \sin \theta$, $y = a \sin \theta - a\theta \cos \theta$.

3. A projectile is fired from the origin with an initial velocity of v_0 in a direction making an angle α with the X-axis. Assuming that the initial impulse and gravity are the only forces operating, find the parametric equations of the path, using time t (in seconds) as parameter.

Ans. $x = v_0 t \cos \alpha$, $y = v_0 t \sin \alpha - \frac{g}{2} t^2$. Eliminate t and show that the path is the parabola $y = x \tan \alpha - \frac{gx^2}{2 v_0^2} \sec^2 \alpha$.

4. Generalize the cycloid (Fig. 164) by letting P be on a radius CP of the rolling circle at a distance R from the center C. Show that the parametric equations of the curve traced by P (general name, a *trochoid; curtate cycloid* if $PC < a$, *prolate cycloid* if $PC > a$) are $x = a\theta - R \sin \theta$, $y = a - R \cos \theta$.

CHAPTER XVII

EMPIRICAL EQUATIONS

91. Curve Fitting. So far we have been concerned with graphs of functions which were exactly expressible in some analytic form. However. much of the work in the sciences relates to data that are functionally related by no known exact formula. In some cases theoretical considerations dictate the form of the function, in others the data themselves suggest a possible relation; but in many cases there is no hint as to what the relation might be.

In any event it is desirable to know the form of the exact functional relation, if one exists, or some approximation which will adequately describe the relation for matters of interpolation, extrapolation, etc. The process of finding the equation of a curve which passes through or near the points of a set of paired observations is called *curve fitting*, and the equation of the curve thus fitted is called an *empirical equation.*

In curve fitting we seek the simplest curve which will reasonably explain the data. We treat four types: *linear*, *parabolic*, *exponential*, and *power*.

92. The Linear Law. Let the observed data be

x_1	x_2	x_3	$\cdots$	x_n
y_1	y_2	y_3	$\cdots$	y_n

When plotted on rectangular coordinate paper these points may appear to lie almost on a straight line. A line could be drawn, by sight, that might suffice for rough work, but there are two standard methods of fitting a line to these observations which eliminate the guesswork. These are I, the *method of averages* and II, the *method of least squares.*

I. *The Method of Averages.* This process involves dividing the given data into two sets equal or nearly equal in numbers of observations, computing the *average point* for each set, and finding the equation of the line passing through these two average points. If the x's in the first set run from x_1 to x_j the average x for that set is

$$\bar{x}_1 = \frac{x_1 + x_2 + \cdots + x_j}{j}.$$

The average points are therefore given by $(\bar{x}_1, \bar{y}_1)$ and $(\bar{x}_2, \bar{y}_2)$ where

$$\bar{x}_2 = \frac{x_{j+1} + x_{j+2} + \cdots + x_n}{n - j},$$

and similarly for the y's.

Illustration 1. By the method of averages fit a straight line to the following data.

x	1	3	5	7	10
y	2	4	6	7	10

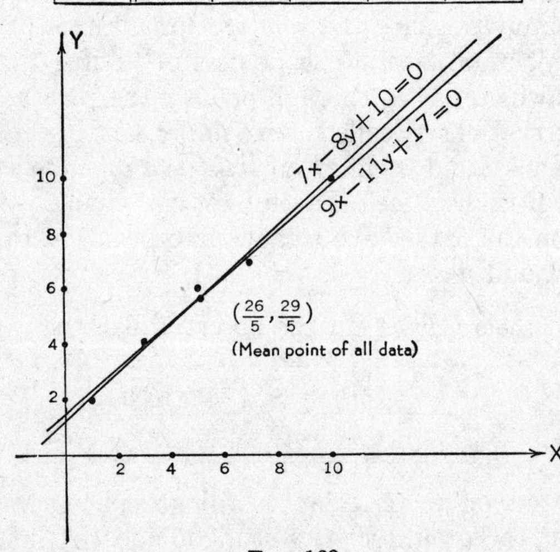

FIG. 169

Solution. The plotted points show a linear trend. We put the first three pairs in the first set and obtain as the average point of this set

$$\bar{x}_1 = \frac{1 + 3 + 5}{3} = 3,$$

$$\bar{y}_1 = \frac{2 + 4 + 6}{3} = 4.$$

Similarly the average point of the second set is

$$\bar{x}_2 = \frac{7 + 10}{2} = \frac{17}{2},$$

$$\bar{y}_2 = \frac{7 + 10}{2} = \frac{17}{2}.$$

The equation of the fitted line passing through (3, 4) and $(\frac{17}{2}, \frac{17}{2})$ is $9x - 11y + 17 = 0$.

If we put only the first two pairs in the first set and the last three pairs in the second set the equation of the line would be $7x - 8y + 10 = 0$. With different groupings of the data this method therefore yields different results.

A point on the fitted line $y_T = a + bx$ is called an estimated or *theoretical point*. The subscript T is used on y to indicate this. An observed value of y (from the data) is often written y_O. For a given abscissa the difference $E = y_O - y_T$ is called the residual or *error*. Even though the line fitted by the method of averages depends upon the grouping, the method yields a line for which *the sum of the errors is zero* and each such line passes through the mean point of all the data. It is customary in mathematical writing to indicate "the sum of" by the Greek capital Σ. Thus "the sum of the errors is zero" is written $\Sigma E = 0$. We note that the condition $\Sigma E = 0$ is not a good criterion for the goodness of fit.

II. *The Method of Least Squares.* A better criterion for goodness of fit, indeed the *best* in a certain probability sense, is that *the sum of squares of the errors shall be a minimum;* hence the name "least squares." This method leads to a unique line generally called the *line of regression* (of y on x). The coefficients a and b in the equation of the regression line must satisfy the system of *normal equations*

$$an + b\Sigma x = \Sigma y, \quad (1)$$

$$a\Sigma x + b\Sigma x^2 = \Sigma xy. \quad (2)$$

To solve these simultaneously for a and b, thus obtaining the line of best fit $y = a + bx$, we first prepare the following table

x_1	y_1	x_1^2	x_1y_1
x_2	y_2	x_2^2	x_2y_2
.	.	.	.
x_n	y_n	x_n^2	x_ny_n
Σx	Σy	Σx^2	Σxy

Illustration 2. By least squares fit a trend line to the data in Illustration 1.

Solution.

x	y	x^2	xy
1	2	1	2
3	4	9	12
5	6	25	30
7	7	49	49
10	10	100	100
$\Sigma x = 26$	$\Sigma y = 29$	$\Sigma x^2 = 184$	$\Sigma xy = 193$

The normal equations are

$$5a + 26b = 29,$$
$$26a + 184b = 193.$$

These yield $a = 318/244$, $b = 211/244$, and the line $y = 318/244 + 211x/244$ or $211x - 244y + 318 = 0$.

93. The Parabolic Law. The observed data might quite obviously be non-linear, in which case it would be undesirable to fit a straight line. In some instances a parabola of the form $y = a + bx + cx^2$ is used. The method of averages would apply but the data would have to be divided into three sets as there are now three coefficients, a, b, and c, to be determined. However, we shall illustrate with the generally applicable method of least squares since it gives better results. The work is longer; there are three normal equations in this case, namely

$$an + b\Sigma x + c\Sigma x^2 = \Sigma y,$$
$$a\Sigma x + b\Sigma x^2 + c\Sigma x^3 = \Sigma xy,$$
$$a\Sigma x^2 + b\Sigma x^3 + c\Sigma x^4 = \Sigma x^2y.$$

Illustration. By least squares fit a parabola to the data

x	0	1	2	3	4
y	1	2	2.5	3	3

Solution. We prepare the table

x	y	x^2	x^3	x^4	xy	x^2y
0	1	0	0	0	0	0
1	2	1	1	1	2	2
2	2.5	4	8	16	5	10
3	3	9	27	81	9	27
4	3	16	64	256	12	48
$\Sigma x=10$	$\Sigma y=11.5$	$\Sigma x^2=30$	$\Sigma x^3=100$	$\Sigma x^4=354$	$\Sigma xy=28$	$\Sigma x^2y=87$

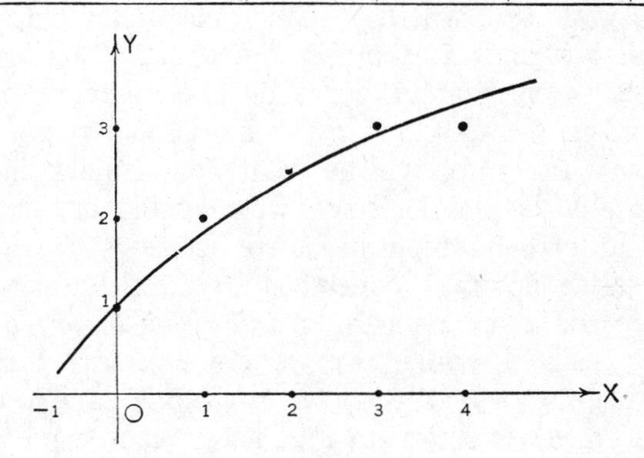

FIG. 170

The normal equations are

$$5\,a + 10\,b + 30\,c = 11.5,$$
$$5\,a + 15\,b + 50\,c = 14 \text{ (reduced)},$$
$$30\,a + 100\,b + 354\,c = 87.$$

These yield $a = 7.1/7$, $b = 7.5/7$, $c = -1/7$; and the parabolic equation is

$$7\,y = 7.1 + 7.5\,x - x^2.$$

94. The Exponential Law. Any quantity that increases at a given time x at a rate proportional to the amount present at that time is said to increase exponentially, and the amount y present at any time x is given by

$$y = ae^{bx} \tag{1}$$

where a and b are constants. If b is negative the quantity is decreasing.

Money continuously compounded follows this law, and it is otherwise important in biology, chemistry, etc., where the quantity referred to might be certain organisms or certain chemicals. Taking logarithms to base e on both sides of (1) we get

$$\log y = \log a + bx, \tag{2}$$

which shows that if the exponential law obtains then the points $(x, \log y)$ fall on a straight line since (2) is a linear equation in x and $\log y$. The results are similar regardless of the base of logarithms used.

It is easy to test whether given data follow (nearly) an exponential law and two courses are open: (a) to compute $\log y$ for each y entry and plot $(x, \log y)$ on *rectangular* coordinate paper, or (b) to plot (x, y) on *semi-logarithmic* paper. In either case the graph will be (nearly) a straight line. The latter course is the simpler since, with semi-logarithmic paper available, no further computations are necessary.

Semi-logarithmic graph paper has the usual linear scale one way, where the mark x means *x units from the origin,* but a logarithmic scale the other way, where the mark y is placed at a distance *$\log y$ units from the origin.* A sample of semi-logarithmic paper is shown in Fig. 171. Since $\log 1 = 0$, the origin is marked unity; there is no zero on a logarithmic scale.

If we set, in (2), $\log y = Y$ and $\log a = A$, we get

$$Y = A + bx. \tag{3}$$

Fitting a straight line (3) to given data will readily yield the exponential law $y = ae^{bx}$. If it is more convenient to use logarithms to the base 10 we write

$$\begin{aligned} Y' = \log_{10} y &= \log_{10} a + bx \log_{10} e \\ &= A' + .43429\, bx \end{aligned}$$

or

$$Y' = A' + b'x \tag{4}$$

and proceed with this equation.

Illustration. Fit a curve of the exponential type $y = ae^{bx}$ to the data

x	1	2	3	4
y	8	5	3	2

Solution. We use (3) and logarithms to the base *e* (Table III, Appendix B) to prepare the following table.

x	y	$Y = \log y$	x^2	xY
1	8	2.0794	1	2.0794
2	5	1.6094	4	3.2188
3	3	1.0986	9	3.2958
4	2	0.6932	16	2.7728
$\Sigma x = 10$		$\Sigma Y = 5.4806$	$\Sigma x^2 = 30$	$\Sigma xY = 11.3668$

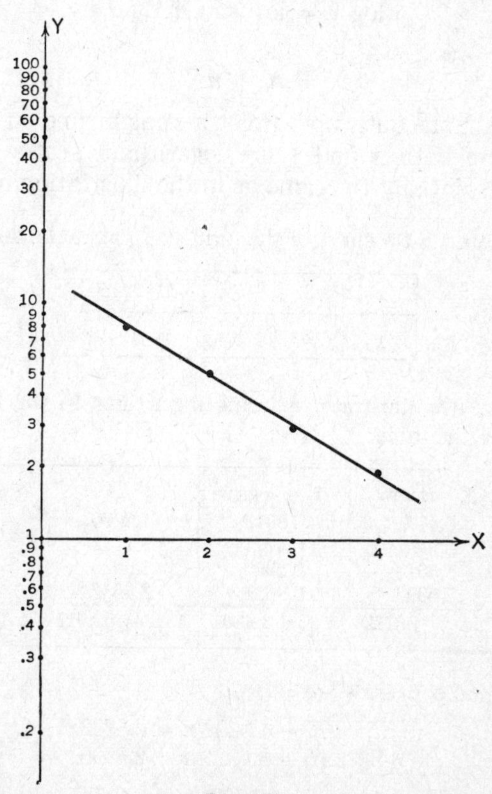

FIG. 171

The normal equations are

$$2A + 5b = 2.7403 \text{ (reduced)},$$
$$5A + 15b = 5.6834 \text{ (reduced)},$$

yielding $A = 2.5378$, $b = -0.4669$. Equation (3) becomes

$$Y = 2.5378 - 0.4669\,x.$$

Since $A = \log a = 2.5378$, $a = 12.62$ (from a larger table of natural logarithms).

Hence the fitted exponential curve has the equation

$$y = 12.62\,e^{-0.4669\,x}.$$

95. The Power Law. Data might follow the power law $y = kx^r$. Taking logarithms on both sides we get

$$\log y = \log k + n \log x,$$

or

$$(1) \qquad Y = K + nX$$

showing that the data would plot a straight line on *logarithmic paper* where both x and y are logarithmic scales. The procedure is essentially the same as in the illustration of § 94.

Illustration. Fit a curve of the form $y = kx^n$ to the data

x	3	7	20	50
y	2	3	5	8

Solution. We illustrate by using logarithms to the base 10 and prepare the table:

x	y	$X = \log_{10} x$	$Y = \log_{10} y$	X^2	XY
3	2	0.4771	0.3010	.2276	.1426
7	3	0.8451	0.4771	.7142	.4032
20	5	1.3010	0.6990	1.6926	.8094
50	8	1.6990	0.9031	2.8567	1.5344
		$\Sigma X = 4.3222$	$\Sigma Y = 2.3802$	$\Sigma X^2 = 5.4911$	$\Sigma XY = 2.8896$

The normal equations are

$$4K + 4.3222\,n = 2.3802,$$
$$4.3222\,K + 5.4911\,n = 2.8896,$$

whose solution is $\quad K = .1768.\ n = .3871.$

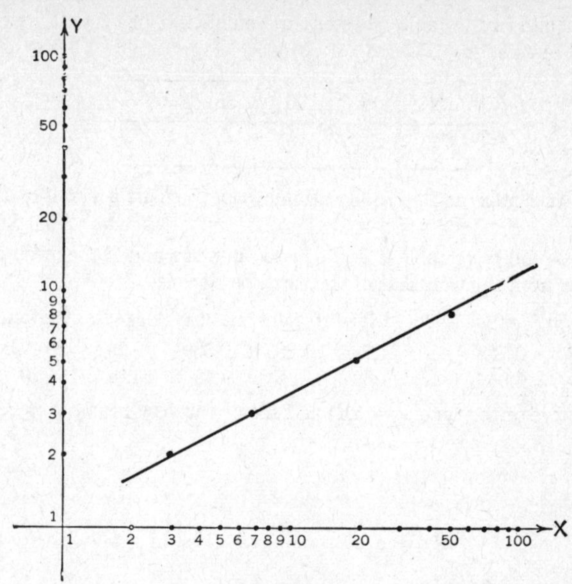

FIG. 172

Now

$$K = \log k = .1768,$$
$$k = 1.502.$$

The fitted curve, therefore, has the equation

$$y = 1.502\, x^{0.3871}.$$

EXERCISES

(Use method of least squares in each case.)

1. Fit a straight line to the data

x	-2	-1	0	1	2
y	-2	-1	.5	1	1.5

Ans. $y = .9\,x$.

2. Fit a parabola to the data

x	0	1	2	3
y	3	1	1	2

Ans. $20\,y = 59 - 51\,x + 15\,x^2$.

3. The data in the table represent the number N of bacteria in a culture at the end of t hours.

N	10	20	35	65
t	1	2	3	4

Plot on rectangular and semi-logarithmic paper and fit a curve of the type $N = ae^{bt}$. *Ans.* $N = 5.55\,e^{0.62\,t}$.

4. In the following table T is the period in years and R the mean distance of a planet from the sun, that of the earth being unity.

	Mercury	Venus	Earth	Mars	Jupiter	Saturn	Uranus	Neptune	Pluto
R	0.387	0.723	1	1.524	5.203	9.539	19.191	30.071	39.5
T	0.24	0.61	1	1.88	11.86	29.46	84.02	164.79	248

Fit a curve of the type $T = kR^n$ and show that very nearly $T = R^{\frac{3}{2}}$.

SOLID ANALYTIC GEOMETRY

CHAPTER XVIII

FUNDAMENTAL CONCEPTS

96. Coordinate Systems. Three standard methods are used to locate points in three dimensions: I, *Rectangular Coordinates;* II, *Cylindrical Coordinates;* and III, *Spherical Coordinates.*

I. *Rectangular Coordinates.* The point $P(x, y, z)$ is located by its distances x, y, and z from three mutually perpendicular *coordinate planes,* the YZ-, XZ-, and XY-planes respectively. The lines of intersection of these planes are called the *coordinate axes.* In Fig. 173 the XZ-plane coincides with the plane of the

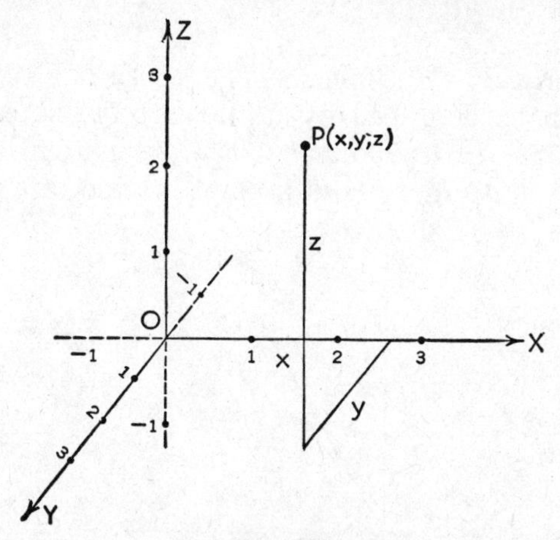

FIG. 173

paper and the same distance is used for the unit on the x- and z-axes. A lesser distance is used for the unit on the y-axis since there is foreshortening. However, the axes may be drawn and labeled in whatever way seems most desirable for a particular problem. The axes in Fig. 173 form a left-handed system; if the X- and Y-axes are interchanged a right-handed system is obtained. For special reasons a right-handed system is generally used with problems in vector mechanics. In rectangular coordinates in space there is a one-to-one correspondence between points and number triplets.

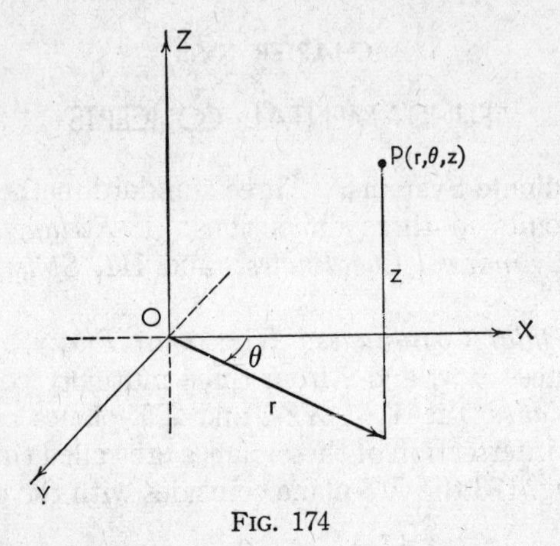

FIG. 174

II. *Cylindrical Coordinates*. The position of a point $P(r, \theta, z)$ can be described by its polar coordinates, r and θ, in the XY-plane, and the rectangular coordinate z. (Fig. 174). The cylindrical coordinates of a point in space are not unique

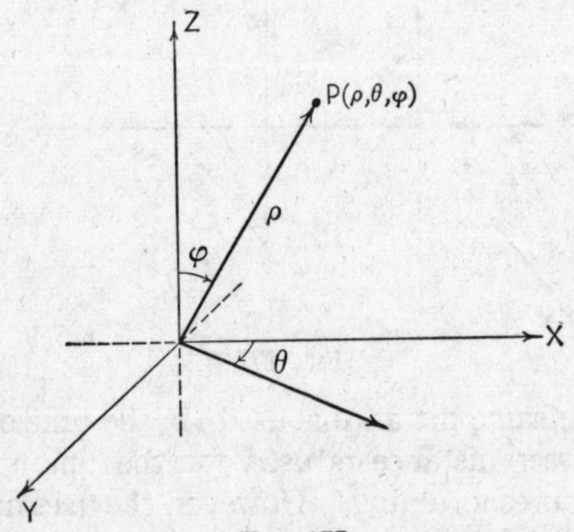

FIG. 175

but despite this drawback the system is very useful in many problems.

III. *Spherical Coordinates.* The point $P(\rho, \theta, \phi)$ can be located by the spherical coordinates ρ, θ, and ϕ as indicated in Fig. 175. Again the coordinates of a point in this system are not unique but many problems are readily treated by spherical coordinates and are difficult in rectangular or cylindrical coordinates.

The following substitutions will transform from one system to another.

Rectangular to Cylindrical

$$x = r\cos\theta$$
$$y = r\sin\theta$$
$$z = z$$

Cylindrical to Rectangular

$$r = \sqrt{x^2 + y^2}$$
$$\theta = \tan^{-1}\frac{y}{x}$$
$$z = z$$

Rectangular to Spherical

$$x = \rho\sin\phi\cos\theta$$
$$y = \rho\sin\phi\sin\theta$$
$$z = \rho\cos\phi$$

Spherical to Rectangular

$$\rho = \sqrt{x^2 + y^2 + z^2}$$
$$\theta = \tan^{-1}\frac{y}{x}$$
$$\phi = \cos^{-1}\frac{z}{\sqrt{x^2 + y^2 + z^2}}$$

Cylindrical to Spherical

$$r = \rho\sin\phi$$
$$\theta = \theta$$
$$z = \rho\cos\phi$$

Spherical to Cylindrical

$$\rho = \sqrt{r^2 + z^2}$$
$$\theta = \theta$$
$$\phi = \tan^{-1}\frac{r}{z}$$

The coordinate planes divide space into eight *octants*, the *first octant* being the region where x, y, and z are positive. In general the other octants are not labeled. The majority of our work will be in rectangular coordinates.

97. Distance between Two Points. From Fig. 176 and the formula for distance between two points in a plane we have

$$\begin{aligned}\overline{P_1P_2}^2 &= \overline{P_1A}^2 + \overline{AP_2}^2\\ &= (x_2 - x_1)^2 + (y_2 - y_1)^2 + (z_2 - z_1)^2,\end{aligned}$$

or

$$d = \sqrt{(x_2 - x_1)^2 + (y_2 - y_1)^2 + (z_2 - z_1)^2}.$$

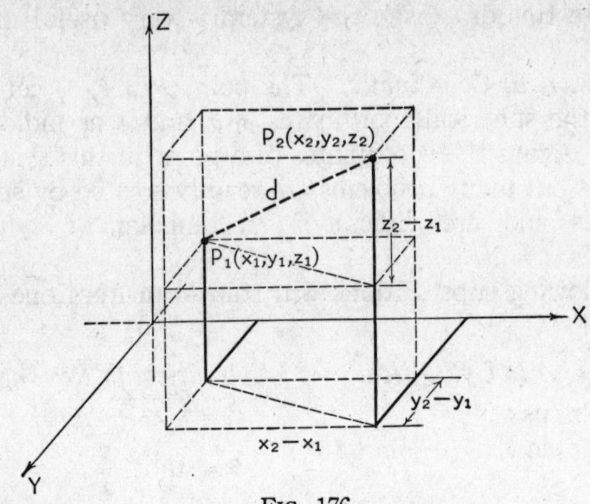

FIG. 176

98. Projections. The projection of a signed segment AB (Fig. 177) upon a plane π is the signed segment $A'B'$, lying in the plane, joining the feet of the perpendiculars from A and B to π. Similarly the projection of AB upon a line L is $A''B''$, the signed segment cut off by the perpendiculars dropped upon L from A and B.

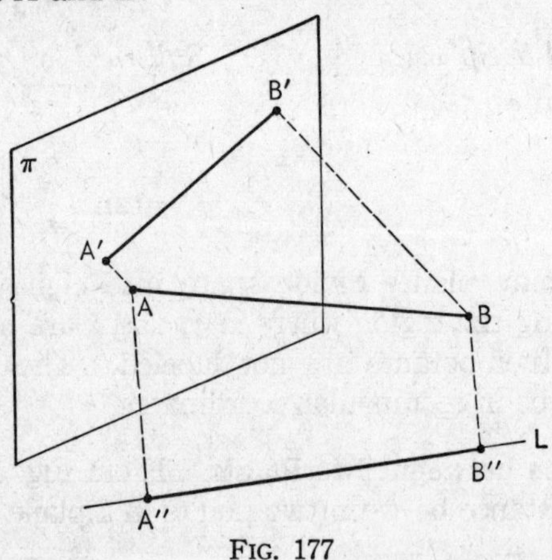

FIG. 177

99. Point of Division. By analogy with the methods of § 10 we find that a point $P(x, y, z)$ will divide the segment P_1P_2 in the ratio r_1/r_2 if

$$(1) \qquad x = \frac{x_1r_2 + x_2r_1}{r_1 + r_2}, \quad y = \frac{y_1r_2 + y_2r_1}{r_1 + r_2}, \quad z = \frac{z_1r_2 + z_2r_1}{r_1 + r_2}.$$

If r_1/r_2 is positive, P is an internal point of division; if r_1/r_2 is negative, P is an external point of division. In particular if P is the midpoint of P_1P_2, then

$$(2) \qquad \bar{x} = \frac{x_1 + x_2}{2}, \quad \bar{y} = \frac{y_1 + y_2}{2}, \quad \bar{z} = \frac{z_1 + z_2}{2}.$$

Equations (1) give the coordinates of the center of gravity of masses r_2 and r_1 placed respectively at P_1 and P_2.

100. Direction Cosines. Two intersecting lines in space determine a plane and the angle between them is defined as in the plane case. Lines which do not intersect are called *skew lines* and the angle between two skew lines is defined as the angle between two intersecting lines parallel to them and having the same sense.

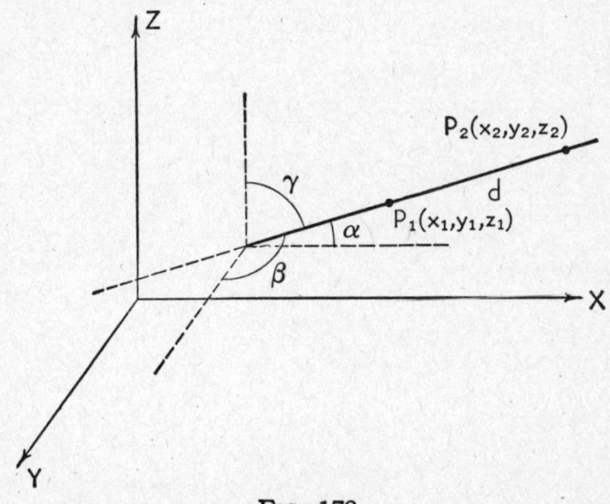

FIG. 178

The angles α, β, and γ which a directed line (or line segment) makes with the positive X-, Y-, and Z-axes respectively are called the *direction angles*. The direction angles are unique for a given sensed line. The *direction cosines* of a sensed line are

$$(1) \qquad \begin{aligned} \lambda &= \cos \alpha, \\ \mu &= \cos \beta, \\ \nu &= \cos \gamma, \end{aligned}$$

while those of a line without established direction are λ, μ, ν or $-\lambda$, $-\mu$, $-\nu$. *Direction numbers* of a line are any numbers a, b, c proportional to the direction cosines. Thus $a = k\lambda$, $b = k\mu$, $c = k\nu$ are direction numbers.

$$(2) \qquad \cos \alpha = \frac{x_2 - x_1}{d}, \ \cos \beta = \frac{y_2 - y_1}{d}, \ \cos \gamma = \frac{z_2 - z_1}{d}.$$

Thus the difference in the respective coordinates of any two points on a line are direction numbers of the line. The constant of proportionality d is the distance between the two points.

Upon squaring and adding (2) we get

$$\lambda^2 + \mu^2 + \nu^2 = 1.$$

Hence

$$\lambda = \frac{a}{\sqrt{a^2 + b^2 + c^2}}, \ \mu = \frac{b}{\sqrt{a^2 + b^2 + c^2}}, \ \nu = \frac{c}{\sqrt{a^2 + b^2 + c^2}}.$$

101. Angle between Two Lines. There is no loss in generality in assuming that the two lines pass through the origin. Let P_1 and P_2 be any points on the lines L_1 and L_2 respectively. Then, by the law of cosines,

$$\overline{P_1P_2}^2 = d_1^2 + d_2^2 - 2\,d_1d_2 \cos \theta,$$

$$(x_2 - x_1)^2 + (y_2 - y_1)^2 + (z_2 - z_1)^2 = x_1^2 + y_1^2 + z_1^2 + x_2^2 + y_2^2 + z_2^2 - 2\,d_1d_2 \cos \theta,$$

$$\cos \theta = \frac{x_1x_2 + y_1y_2 + z_1z_2}{d_1d_2}.$$

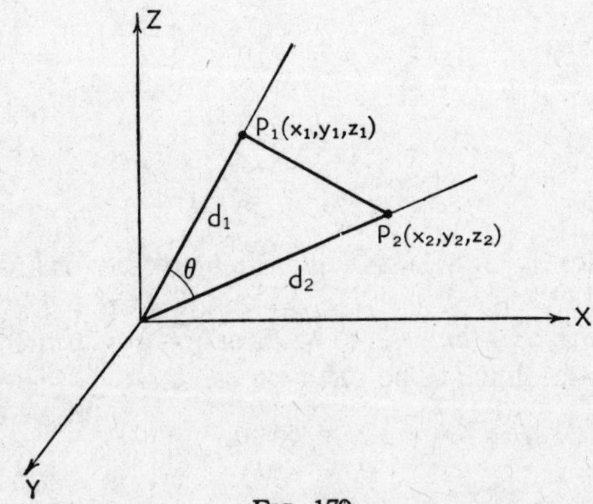

FIG. 179

But

$$\lambda_1 = \frac{x_1}{d_1}, \ \mu_1 = \frac{y_1}{d_1}, \ \nu_1 = \frac{z_1}{d_1},$$

$$\lambda_2 = \frac{x_2}{d_2}, \ \mu_2 = \frac{y_2}{d_2}, \ \nu_2 = \frac{z_2}{d_2}.$$

Therefore the angle θ between two directed lines will be given by

(1) $$\cos \theta = \lambda_1\lambda_2 + \mu_1\mu_2 + \nu_1\nu_2.$$

The acute angle between two undirected lines is given by

(2) $\cos \theta = | \lambda_1\lambda_2 + \mu_1\mu_2 + \nu_1\nu_2 |$. [Compare with (2) and (3), § 13.]

102. Parallel and Perpendicular Lines. Two lines are parallel

(1) If direction cosines are equal:

$$\lambda_1 = \lambda_2, \ \mu_1 = \mu_2, \ \nu_1 = \nu_2;$$

(2) Or if direction numbers are proportional:

$$a_1 = Ka_2, \ b_1 = Kb_2, \ c_1 = Kc_2.$$

Two parallel lines determine a plane; parallelism is essentially a two-dimensional concept.

Two lines are perpendicular

(3) If $\lambda_1\lambda_2 + \mu_1\mu_2 + \nu_1\nu_2 = 0;$

(4) Or if $a_1a_2 + b_1b_2 + c_1c_2 = 0.$

Two perpendicular lines may intersect (determine a plane) or they may be skew lines in space.

Illustration 1. Transform the equation $z^2 = a^2 - 2x^2 - 2y^2$ to cylindrical coordinates.

Solution. Since $r^2 = x^2 + y^2$

the cylindrical equation is

$$z^2 = a^2 - 2r^2.$$

Illustration 2. Find the point P on $P_1(2, 3, 5)$, $P_2(-1, 2, -3)$ which divides P_1P_2 in the ratio $\frac{2}{3}$.

Solution.

$$x = \frac{(2)(3)+(-1)(2)}{5}, \ y = \frac{(3)(3)+(2)(2)}{5}, \ z = \frac{(5)(3)+(-3)(2)}{5}.$$

The point is $P(\frac{4}{5}, \frac{13}{5}, \frac{9}{5})$.

Illustration 3. Find the angle between the lines joining $P(1, 2, -3)$, $Q(-1, -2, 1)$, and $R(0, 1, 4)$, $S(1, -3, 0)$.

Solution.

$$PQ = \sqrt{4 + 16 + 16} = 6,$$
$$RS = \sqrt{1 + 16 + 16} = \sqrt{33},$$
$$\lambda_1 = -\tfrac{1}{3}, \quad \mu_1 = -\tfrac{2}{3}, \quad \nu_1 = \tfrac{2}{3},$$
$$\lambda_2 = -\frac{1}{\sqrt{33}}, \quad \mu_2 = \frac{4}{\sqrt{33}}, \quad \nu_2 = \frac{4}{\sqrt{33}}.$$
$$\cos\theta = \frac{1 - 8 + 8}{3\sqrt{33}} = \frac{1}{99}\sqrt{33}.$$

Illustration 4. Determine c_2 so that the two lines $L_1 : a_1 = 1$, $b_1 = -1$, $c_1 = 2$; $L_2 : a_2 = 2$, $b_2 = 4$, $c_2 = ?$, are perpendicular.

Solution.

$$a_1a_2 + b_1b_2 + c_1c_2 = 0,$$
$$2 - 4 + 2\,c_2 = 0,$$
$$c_2 = 1.$$

EXERCISES

1. Transform the equation $\rho = 2$ to rectangular coordinates.

Ans. $x^2 + y^2 + z^2 = 4$.

2. Find the midpoint of the segment $P(-2, 3, -4)$, $Q(8, 5, 2)$.

Ans. $(3, 4, -1)$.

3. Find the direction cosines of the line making equal angles with the axes.

Ans. $\lambda = \mu = \nu = \frac{1}{3}\sqrt{3}$.

4. Show that the line through $(5, 1, -2)$ and $(-4, -5, 13)$ is the perpendicular bisector of the segment $(-5, 2, 0)$, $(9, -4, 6)$.

CHAPTER XIX

THE PLANE

103. Equations in Three Variables. We have seen that an equation in two variables, $f(x, y) = 0$, plots a curve in two-space. An equation in three variables, $f(x, y, z) = 0$, plots a *surface* in three-space.

Two equations $f(x, y) = 0$, $g(x, y) = 0$, solved simultaneously, give the points of intersection of the two curves. Similarly two equations $f(x, y, z) = 0$, $g(x, y, z) = 0$, solved simultaneously, give the points of intersection of the two surfaces; these points generally lie on a space curve which is the curve of intersection of the surfaces. The simultaneous solution of three equations $f(x, y, z) = 0$, $g(x, y, z) = 0$, $h(x, y, z) = 0$ yields the points common to the three surfaces. These points are, in general, isolated points.

It is important to note that a curve in space is represented by *two* simultaneous equations.

104. The Linear Equation. The simplest equation in three variables is the linear equation

$$Ax + By + Cz + D = 0. \tag{1}$$

The left-hand member is a polynomial of the first degree in the three variables x, y, and z. We prove the following two theorems.

Theorem 1. The graph of a linear equation in three variables is a plane.

Proof. Let $P_1(x_1, y_1, z_1)$ and $P_2(x_2, y_2, z_2)$ be two points on the surface

$$Ax + By + Cz + D = 0.$$

It follows that

$$Ax_1 + By_1 + Cz_1 + D = 0, \tag{2}$$

$$Ax_2 + By_2 + Cz_2 + D = 0. \tag{3}$$

Multiplying (2) by $r_2/(r_1 + r_2)$ and (3) by $r_1/(r_1 + r_2)$ and adding we get

$$A\frac{x_1r_2 + x_2r_1}{r_1 + r_2} + B\frac{y_1r_2 + y_2r_1}{r_1 + r_2} + C\frac{z_1r_2 + z_2r_1}{r_1 + r_2} + D = 0,$$

which shows that *any* point on the line P_1P_2 lies on (1). (See § 99.) The plane is the only surface such that the line joining any two points on it lies wholly in it. Hence (1) is a plane.

Theorem 2. A plane is represented by a linear equation.

Proof. Consider a point $P_1(x_1, y_1, z_1)$ on the plane π and the normal line L to π at P_1. Let direction numbers of L be A, B, C. If $P(x, y, z)$ is any other point in π then L and P_1P are perpendicular to each other. Direction numbers of P_1P are $(x - x_1), (y - y_1), (z - z_1)$. By (4), § 102, we have

$$A(x - x_1) + B(y - y_1) + C(z - z_1) = 0. \tag{4}$$

This holds for all points P in π and no others; hence (4) is the equation of π. Since (4) is linear the theorem is proved.

The equations of the coordinate planes are $x = 0$, $y = 0$, $z = 0$, the YZ-, XZ-, XY-planes respectively. The points of intersection of a plane with the axes are called the *intercepts* of the plane. The x-intercept is located by setting $y = z = 0$ in the equation of the plane and solving for x. Thus the x-intercept is $(-D/A, 0, 0)$. The other intercepts are similarly determined. The lines of intersection with the coordinate planes are called the *traces* of the plane. The trace in the XY-plane is given by the simultaneous equations

$$\begin{aligned} Ax + By + Cz + D &= 0, \\ z &= 0. \end{aligned}$$

Hence *in* the XY-plane the equation of the trace is $Ax + By + D = 0$. The other traces are determined in like manner.

The plane is the space analogue of the line and the student should review the latter at this time (Chapter IV).

105. Special Forms of the Equation of a Plane. The *general* equation, $Ax + By + Cz + D = 0$, can be written in the following special forms, which are both useful and interesting.

I. *Three-Point Form.* A plane is determined by three non-collinear points $P_1(x_1, y_1, z_1)$, $P_2(x_2, y_2, z_2)$, $P_3(x_3, y_3, z_3)$. These must satisfy the general equation of the plane; that is,

$$(1) \qquad Ax_1 + By_1 + Cz_1 + D = 0,$$
$$(2) \qquad Ax_2 + By_2 + Cz_2 + D = 0,$$
$$(3) \qquad Ax_3 + By_3 + Cz_3 + D = 0.$$

These three equations can be solved for the coefficients A, B, C, D (only three of them are effective). When these coefficients are substituted into $Ax + By + Cz + D = 0$ we have the three-point form of the equation of a plane. The result can be written most readily in determinant form:

$$(4) \qquad \begin{vmatrix} x & y & z & 1 \\ x_1 & y_1 & z_1 & 1 \\ x_2 & y_2 & z_2 & 1 \\ x_3 & y_3 & z_3 & 1 \end{vmatrix} = 0.$$

This can be written in semi-expanded form, using determinants of the third order:

$$(5) \quad \begin{vmatrix} y_1 & z_1 & 1 \\ y_2 & z_2 & 1 \\ y_3 & z_3 & 1 \end{vmatrix} x - \begin{vmatrix} x_1 & z_1 & 1 \\ x_2 & z_2 & 1 \\ x_3 & z_3 & 1 \end{vmatrix} y + \begin{vmatrix} x_1 & y_1 & 1 \\ x_2 & y_2 & 1 \\ x_3 & y_3 & 1 \end{vmatrix} z - \begin{vmatrix} x_1 & y_1 & z_1 \\ x_2 & y_2 & z_2 \\ x_3 & y_3 & z_3 \end{vmatrix} = 0.$$

II. *Point-Direction Form.* We have already developed this form of the equation of a plane in the proof of theorem 2, § 104. In (4) of that paragraph we wrote the point-direction form:

$$A(x - x_1) + B(y - y_1) + C(z - z_1) = 0.$$

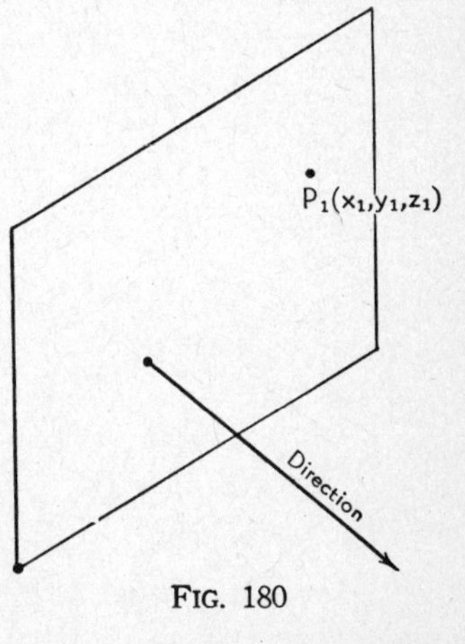

FIG. 180

The student should make sure that he understands why the coefficients of x, y, and z in the equation of a plane are direction numbers of lines perpendicular to the plane.

III. *Normal Form.* It is easy to convert direction numbers to direction cosines. In $Ax + By + Cz + D = 0$, A, B, C

are direction numbers of lines perpendicular to the plane; if the equation is divided by $\sqrt{A^2 + B^2 + C^2}$ we get

$$(6) \qquad \frac{Ax + By + Cz + D}{\sqrt{A^2 + B^2 + C^2}} = 0.$$

The coefficients of x, y, and z are now the direction cosines of lines perpendicular to the plane. This equation can be written in the form

$$(7) \qquad \lambda x + \mu y + \nu z - p = 0$$

where p is the perpendicular distance from the origin to the plane. For (Fig. 181) let $ON = p$ and let P be any point in

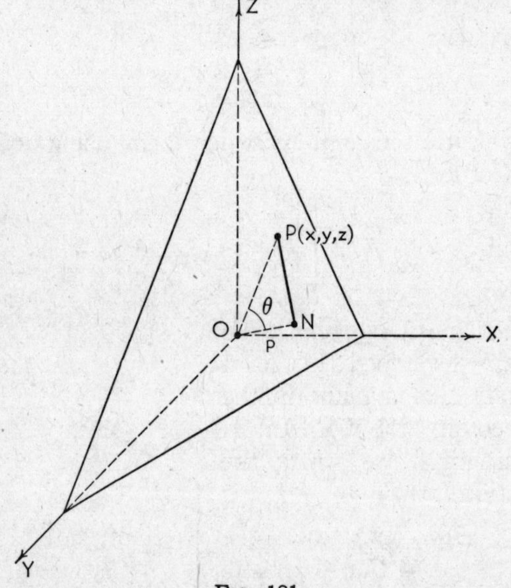

FIG. 181

the plane. The direction cosines of ON are λ, μ, ν and those of OP are x/OP, y/OP, z/OP. Since $p = OP \cos\theta$ we have, by (1), § 101,

$$p = OP\left(\frac{\lambda x}{OP} + \frac{\mu y}{OP} + \frac{\nu z}{OP}\right)$$

which reduces to (7).

Equation (7) is the normal form of the equation of a plane and (6) indicates the method of reducing the general equation to normal form. The sign before the radical $\sqrt{A^2 + B^2 + C^2}$ is chosen so that p is positive. This is done as follows: The sign of the radical is

Opposite to D, if $D \neq 0$;
The same as C, if $D = 0$, $C \neq 0$;
The same as B, if $D = C = 0$, $B \neq 0$;
The same as A, if $D = C = B = 0$, $A \neq 0$.

IV. *Intercept Form.* In case the intercepts are a, b, and c the three-point form reduces to

$$\frac{x}{a} + \frac{y}{b} + \frac{z}{c} = 1. \tag{8}$$

V. *Through Origin.* The general equation reduces to

$$Ax + By + Cz = 0 \tag{9}$$

if the plane passes through (0, 0, 0).

VI. *Perpendicular to Coordinate Planes.* Since $Ax + By + D = 0$ represents, in two-space, a line in the XY-plane, this same equation will represent a plane in three-space perpendicular to the XY-plane. This is true because the equation holds for every point (x, y) on the line regardless of the value of z. Every point (x, y) on the line may be thought of as being projected in the z direction. The projection of a line is, in general, a plane.

The equation of the plane perpendicular to the

XY-plane is $Ax + By + D = 0$;
XZ-plane is $Ax + Cz + D = 0$;
YZ-plane is $By + Cz + D = 0$.

VII. *Perpendicular to Axes.* In one-space $Ax + D = 0$, or $x = k$, represents a point; in two-space it represents a line. In three-space it represents a plane perpendicular to the X-axis.

The equation of the plane perpendicular to the

X-axis is $x = k$;
Y-axis is $y = k$;
Z-axis is $z = k$.

Illustration 1. Find the equation of the plane passing through the three points (1, 2, 1), (−1, 1, −2), (1, 0, −1).

Solution. Equation (5) becomes

$$\begin{vmatrix} 2 & 1 & 1 \\ 1 & -2 & 1 \\ 0 & -1 & 1 \end{vmatrix} x - \begin{vmatrix} 1 & 1 & 1 \\ -1 & -2 & 1 \\ 1 & -1 & 1 \end{vmatrix} y + \begin{vmatrix} 1 & 2 & 1 \\ -1 & 1 & 1 \\ 1 & 0 & 1 \end{vmatrix} z - \begin{vmatrix} 1 & 2 & 1 \\ -1 & 1 & -2 \\ 1 & 0 & -1 \end{vmatrix} = 0,$$

which reduces to $x + y - z - 2 = 0$.

Illustration 2. Find the equation of the plane perpendicular to the line joining (1, 2, 0) and (3, −2, 5) and passing through (1, 5, −8).

Solution. Direction numbers of the line joining (1, 2, 0) and (3, −2, 5) are 2, −4, 5. The point-direction form is

$$A(x - x_1) + B(y - y_1) + C(z - z_1) = 0,$$

which becomes

$$2(x - 1) - 4(y - 5) + 5(z + 8) = 0.$$

This can be reduced to

$$2x - 4y + 5z + 58 = 0.$$

Illustration 3. Reduce $3x + 2y - z + 5 = 0$ to normal form.

Solution. The normal form is

$$\frac{3x + 2y - z + 5}{-\sqrt{14}} = 0.$$

Illustration 4. Reduce $x - 5y + 2z - 3 = 0$ to intercept form and write the coordinates of the intercepts.

Solution. The intercept form is

$$\frac{x}{3} + \frac{y}{-\frac{3}{5}} + \frac{z}{\frac{3}{2}} = 1$$

and the intercepts are (3, 0, 0), $(0, -\frac{3}{5}, 0)$ and $(0, 0, \frac{3}{2})$.

Illustration 5. Write the equation of the plane passing through (1, 1, 2) and (3, 5, 4) which is perpendicular to the XY-plane.

Solution. The equation is of the form

$$Ax + By + D = 0.$$

This passes through the given points and hence

$$A + B + D = 0,$$
$$3A + 5B + D = 0,$$

from which we get $A = -2D$, $B = D$.
The equation is, therefore,

$$2x - y - 1 = 0.$$

Illustration 6. Find the equation of the plane perpendicular to the plane $x - y + 2z - 5 = 0$, parallel to the line whose direction cosines are $\frac{1}{5}$, $-\frac{2}{5}$, $2\sqrt{5}/5$, and passing through (5, 0, 1).

Solution. The normal form of the equation will be

$$\lambda x + \mu y + \nu z - p = 0. \tag{1}$$

The direction cosines of the normal to the given plane are $1/\sqrt{6}$, $-1/\sqrt{6}$, $2/\sqrt{6}$. Since the two planes are to be perpendicular their normals must be perpendicular. Hence we must have

$$\frac{\lambda}{\sqrt{6}} - \frac{\mu}{\sqrt{6}} + \frac{2\nu}{\sqrt{6}} = 0,$$

or, simply,

$$\lambda - \mu + 2\nu = 0. \tag{2}$$

Again, the plane must be parallel to the given line; hence the normal to the plane must be normal to the line. Thus

$$\frac{\lambda}{5} - \frac{2\mu}{5} + \frac{2\sqrt{5}\nu}{5} = 0,$$

or

$$\lambda - 2\mu + 2\sqrt{5}\nu = 0. \tag{3}$$

Solving (2) and (3) simultaneously we get

$$\lambda = 2(\sqrt{5} - 2)\nu,$$
$$\mu = 2(\sqrt{5} - 1)\nu.$$

Then (1) becomes

$$2(\sqrt{5} - 2)\nu x + 2(\sqrt{5} - 1)\nu y + \nu z - p = 0$$

and this must be satisfied by (5, 0, 1). Hence

$$10(\sqrt{5} - 2)\nu + \nu - p = 0,$$

or $p = (10\sqrt{5} - 19)\nu$. We set $\nu = 1$ and obtain the final equation of the plane:

$$2(\sqrt{5} - 2)x + 2(\sqrt{5} - 1)y + z - (10\sqrt{5} - 19) = 0.$$

EXERCISES

1. Find the equation of the plane parallel to $3x - 2y + 6z + 5 = 0$ and passing through (1, 4, 1). *Ans.* $3x - 2y + 6z - 1 = 0$.

2. Write the equations of the following planes:

(a) Parallel to the XZ-plane through (1, 4, 6). *Ans.* $y = 4$;

(b) Parallel to the XY-plane and 2 units from it. *Ans.* $z = \pm 2$;

(c) With intercepts 1, −2. −4. *Ans.* $x/1 + y/-2 + z/-4 = 1$.

3. Find the perpendicular bisecting plane of the line segment (2, 5, −3), (0, −4, 2). *Ans.* $2x + 9y - 5z - 9 = 0.$

4. Find the numerical distance from the origin to the plane $7x - y + z - 2 = 0.$ *Ans.* $2/\sqrt{51}.$

106. Distance from a Plane to a Point. Let the distance from the plane $\lambda x + \mu y + \nu z - p = 0$ to the point $P_1(x_1, y_1, z_1)$ be d. The normal form of the parallel plane through P_1 is

$$\lambda x + \mu y + \nu z - (p + d) = 0. \tag{1}$$

Since P_1 is on (1) we have

$$\lambda x_1 + \mu y_1 + \nu z_1 - p - d = 0,$$

whence

$$d = \lambda x_1 + \mu y_1 + \nu z_1 - p \tag{2}$$

is the formula for the distance from a plane to a point.

This distance will be positive if the plane separates the point P and the origin; it will be negative if P and the origin lie on the same side of the plane.

Illustration 1. Find the distance from the plane $x+y-z+2=0$ to the point (0, −2, 3).

Solution. The normal form of the plane is

$$\frac{x + y - z + 2}{-\sqrt{3}} = 0$$

and the distance to (0, −2, 3) is

$$d = \frac{-2 - 3 + 2}{-\sqrt{3}}$$
$$= \sqrt{3}.$$

Illustration 2. Find the equation of all planes tangent to a sphere of radius 3 units centered at the origin.

Solution. Any such plane will be at a distance of 3 units from the origin. Thus the equation is

$$\lambda x + \mu y + \nu z - 3 = 0.$$

This is a two-parameter family of planes.

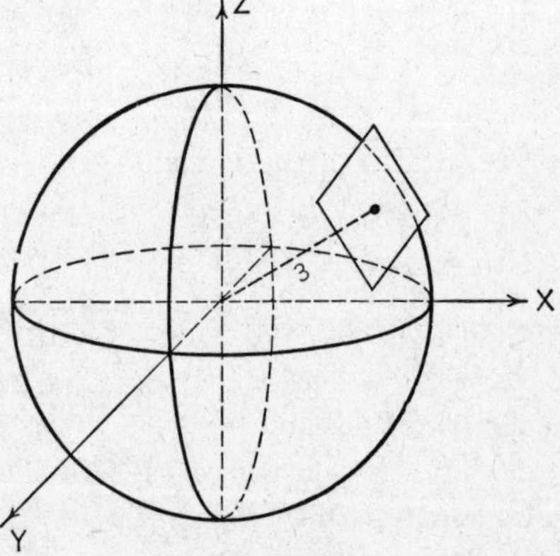

FIG. 182

107. Angle between Two Planes. The angle between two planes is defined as the acute angle between their respective normals. The angle between a line and a plane is defined as the angle between the line and its projection on the plane. Since the coefficients of x, y, and z in the equation of a plane are direction numbers of normals to the plane, the angle between $A_1x + B_1y + C_1z + D_1 = 0$ and $A_2x + B_2y + C_2z + D_2 = 0$ will be given by

$$(1) \qquad \cos\theta = \frac{|A_1A_2 + B_1B_2 + C_1C_2|}{\sqrt{A_1^2 + B_1^2 + C_1^2}\sqrt{A_2^2 + B_2^2 + C_2^2}}.$$

The acute angle between

$$\lambda_1 x + \mu_1 y + \nu_1 z - p_1 = 0$$

and

$$\lambda_2 x + \mu_2 y + \nu_2 z - p_2 = 0$$

will be given by

$$(2) \quad \cos\theta = |\lambda_1\lambda_2 + \mu_1\mu_2 + \nu_1\nu_2|.$$

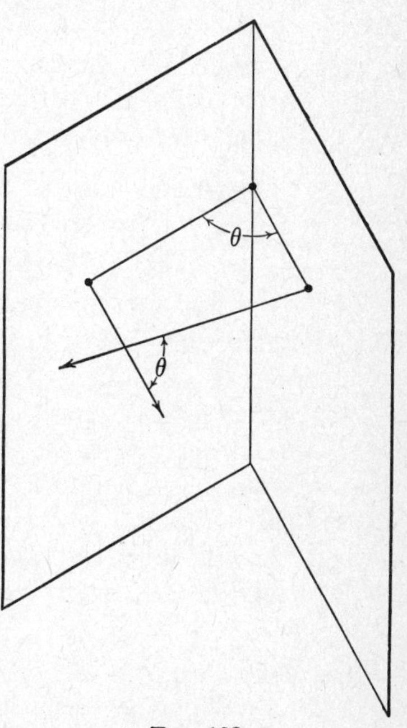

FIG. 183

108. Parallel and Perpendicular Planes. Two planes will be parallel if their normals are parallel. Hence the condition for parallelism is

$$(1) \quad A_1 = KA_2,\ B_1 = KB_2,\ C_1 = KC_2.$$

Two planes will be perpendicular if their normals are perpendicular. Hence the condition for perpendicularity is

$$(2) \quad A_1A_2 + B_1B_2 + C_1C_2 = 0.$$

Illustration 1. Determine the angle between $x + y - z + 7 = 0$ and $3x - y - 6 = 0$.

Solution.

$$\cos\theta = \frac{3 - 1}{\sqrt{3}\sqrt{10}} = \frac{2}{\sqrt{30}}.$$

Illustration 2. Determine the angle between the plane $x + y + 3z - 5 = 0$ and the line joining (1, 2, 3) and (−1, 2, −1).

Solution. The angle between the line and its projection on the plane will be the complement of the angle between the line and the normal to the plane. The direction cosines of the given line are $\sqrt{5}/5, 0, 2\sqrt{5}/5$. Hence the angle sought will be given by

$$\sin\phi = \frac{\sqrt{5} + 6\sqrt{5}}{5\sqrt{11}} = \frac{7\sqrt{5}}{5\sqrt{11}}.$$

109. Systems of Planes. Consider the two planes $\pi_1 \equiv A_1x + B_1y + C_1z + D_1 = 0$ and $\pi_2 \equiv A_2x + B_2y + C_2z + D_2 = 0$. The equation

$$(1)\quad \pi_1 + k\pi_2 \equiv (A_1x + B_1y + C_1z + D_1) + k(A_2x + B_2y + C_2z + D_2) = 0$$

represents a plane since it is linear. It will be satisfied by the coordinates of any point on the line of intersection of $\pi_1 = 0$ and $\pi_2 = 0$ since each parenthesis will then be zero regardless of the value of k. Therefore (1) represents the family of planes through the line of intersection of the two given planes. It is a single-parameter family.

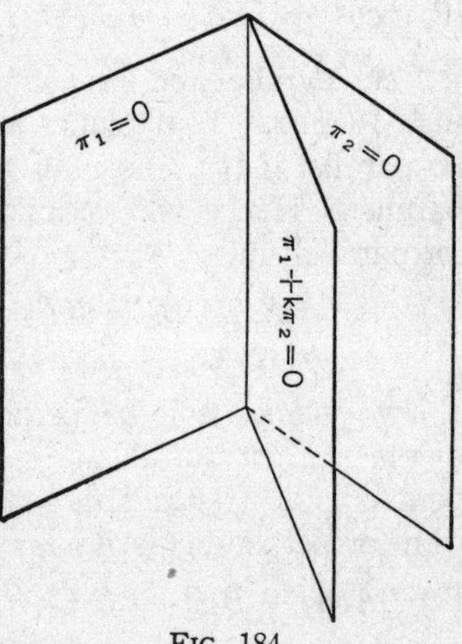

FIG. 184

Illustration. Find the equation of the plane containing the line of intersection of $x+6y-3z+3=0$ and $2x + 7y - z + 8 = 0$ and the point (2, −5, 1).

Solution. The plane will have the equation

$$(x + 6y - 3z + 3) + k(2x + 7y - z + 8) = 0$$

and this must be satisfied by (2, −5, 1). Therefore

$$(2 - 30 - 3 + 3) + k(4 - 35 - 1 + 8) = 0,$$
$$k = -\tfrac{7}{6}.$$

The equation of the plane is

$$(x + 6y - 3z + 3) - \tfrac{7}{6}(2x + 7y - z + 8) = 0,$$
$$8x + 13y + 11z + 38 = 0.$$

110. Condition That Four Planes Be Concurrent. From the theory of determinants it follows that a necessary condition that the four planes

$$A_1x + B_1y + C_1z + D_1 = 0, \qquad A_2x + B_2y + C_2z + D_2 = 0,$$
$$A_3x + B_3y + C_3z + D_3 = 0, \qquad A_4x + B_4y + C_4z + D_4 = 0$$

shall meet in a point is

$$\begin{vmatrix} A_1 & B_1 & C_1 & D_1 \\ A_2 & B_2 & C_2 & D_2 \\ A_3 & B_3 & C_3 & D_3 \\ A_4 & B_4 & C_4 & D_4 \end{vmatrix} = 0.$$

This condition is also sufficient provided no two of the planes are parallel or coincide. This can be written in the following form, using determinants of the third order:

$$A_1\begin{vmatrix} B_2 & C_2 & D_2 \\ B_3 & C_3 & D_3 \\ B_4 & C_4 & D_4 \end{vmatrix} - A_2\begin{vmatrix} B_1 & C_1 & D_1 \\ B_3 & C_3 & D_3 \\ B_4 & C_4 & D_4 \end{vmatrix} + A_3\begin{vmatrix} B_1 & C_1 & D_1 \\ B_2 & C_2 & D_2 \\ B_4 & C_4 & D_4 \end{vmatrix}$$
$$- A_4\begin{vmatrix} B_1 & C_1 & D_1 \\ B_2 & C_2 & D_2 \\ B_3 & C_3 & D_3 \end{vmatrix} = 0.$$

Illustration. Show that the four planes $x - y + z - 1 = 0$, $x+2y-3z+6=0$, $2x-3y+z-1=0$, and $x+3y-2z+5=0$ are concurrent.

Solution. These are distinct, non-parallel planes.

$$\begin{vmatrix} 1 & -1 & 1 & -1 \\ 1 & 2 & -3 & 6 \\ 2 & -3 & 1 & -1 \\ 1 & 3 & -2 & 5 \end{vmatrix} = \begin{vmatrix} 2 & -3 & 6 \\ -3 & 1 & -1 \\ 3 & -2 & 5 \end{vmatrix}$$
$$- \begin{vmatrix} -1 & 1 & -1 \\ -3 & 1 & -1 \\ 3 & -2 & 5 \end{vmatrix} + 2\begin{vmatrix} -1 & 1 & -1 \\ 2 & -3 & 6 \\ 3 & -2 & 5 \end{vmatrix} - \begin{vmatrix} -1 & 1 & -1 \\ 2 & -3 & 6 \\ -3 & 1 & -1 \end{vmatrix}$$
$$= (-12) - (6) + 2(6) - (-6) = 0.$$

111. Condition That Four Points Be Coplanar. A necessary and sufficient condition that the four points $P_1(x_1, y_1, z_1)$, $P_2(x_2, y_2, z_2)$, $P_3(x_3, y_3, z_3)$, $P_4(x_4, y_4, z_4)$, no three of which are collinear, shall lie in a plane is

$$\begin{vmatrix} x_1 & y_1 & z_1 & 1 \\ x_2 & y_2 & z_2 & 1 \\ x_3 & y_3 & z_3 & 1 \\ x_4 & y_4 & z_4 & 1 \end{vmatrix} = 0.$$

This can be written in the following form, using determinants of the third order:

$$x_1 \begin{vmatrix} y_2 & z_2 & 1 \\ y_3 & z_3 & 1 \\ y_4 & z_4 & 1 \end{vmatrix} - x_2 \begin{vmatrix} y_1 & z_1 & 1 \\ y_3 & z_3 & 1 \\ y_4 & z_4 & 1 \end{vmatrix} + x_3 \begin{vmatrix} y_1 & z_1 & 1 \\ y_2 & z_2 & 1 \\ y_4 & z_4 & 1 \end{vmatrix} - x_4 \begin{vmatrix} y_1 & z_1 & 1 \\ y_2 & z_2 & 1 \\ y_3 & z_3 & 1 \end{vmatrix} = 0.$$

Illustration. No three of the four points $(2, 1, 4)$, $(-1, 0, 1)$, $(0, 0, 2)$, $(-1, -1, 1)$ are collinear. Show that they are coplanar.

Solution. No three of the points are collinear, as can be seen by computing direction numbers of the lines joining them in pairs. For example direction numbers of the line joining the first pair are 3, 1, 3 while direction numbers of the line joining the first and third points are 2, 1, 2. Since 3, 1, 3 are not proportional to 2, 1, 2 the first three points do not lie on a line. Similarly for other combinations of the points.

But the four points are coplanar since

$$\begin{vmatrix} 2 & 1 & 4 & 1 \\ -1 & 0 & 1 & 1 \\ 0 & 0 & 2 & 1 \\ -1 & -1 & 1 & 1 \end{vmatrix} = 2 \begin{vmatrix} 0 & 1 & 1 \\ 0 & 2 & 1 \\ -1 & 1 & 1 \end{vmatrix} + \begin{vmatrix} 1 & 4 & 1 \\ 0 & 2 & 1 \\ -1 & 1 & 1 \end{vmatrix} + 0 \begin{vmatrix} 1 & 4 & 1 \\ 0 & 1 & 1 \\ -1 & 1 & 1 \end{vmatrix} + \begin{vmatrix} 1 & 4 & 1 \\ 0 & 1 & 1 \\ 0 & 2 & 1 \end{vmatrix}$$

$$= 2 - 1 + 0 - 1 = 0.$$

112. Résumé of Plane Formulae.

	EQUATION	FORM
(1)	$Ax + By + Cz + D = 0$	General
(2)	$\begin{vmatrix} x & y & z & 1 \\ x_1 & y_1 & z_1 & 1 \\ x_2 & y_2 & z_2 & 1 \\ x_3 & y_3 & z_3 & 1 \end{vmatrix} = 0$	Three-point
(3)	$A(x - x_1) + B(y - y_1) + C(z - z_1) = 0$	Point-direction
(4)	$\lambda x + \mu y + \nu z - p = 0$	Normal

(5)	$\frac{x}{a} + \frac{y}{b} + \frac{z}{c} = 1$	Intercept
(6)	$Ax + By + Cz = 0$	Through origin
		Perpendicular to:
	$Ax + By + D = 0$	XY-plane
(7)	$Ax + Cz + D = 0$	XZ-plane
	$By + Cz + D = 0$	YZ-plane
		Perpendicular to:
	$x = k$	X-axis
(8)	$y = k$	Y-axis
	$z = k$	Z-axis

EXERCISES

1. Find the numerical distance from the plane $3x + 2y + z - 6 = 0$ to the nearer of the two points $(1, -2, 3)$, $(0, 3, -3)$. *Ans.* $\frac{3}{14}\sqrt{14}$.

2. Find the locus of a point which is always $7/\sqrt{3}$ units from $x+y+z+1=0$.
Ans. The two planes $x + y + z - 6 = 0$ and $x + y + z + 8 = 0$.

6. Find the angle between the XY-plane and $7x - 2y + z - 3 = 0$.
Ans. $\cos\theta = \frac{1}{54}\sqrt{54}$.

4. Show that the three points $P_1(1, 1, 1)$, $P_2(2, 3, -4)$, $P_3(-1, -3, 11)$ are collinear. *Ans.* $a_{12} = 1$, $b_{12} = 2$, $c_{12} = -5$; $a_{13} = 2$, $b_{13} = 4$, $c_{13} = -10$.

5. Find the plane containing the origin and the line of intersection of $5x + y - 2z - 8 = 0$ and $4x - 3y + z + 4 = 0$. *Ans.* $13x - 5y = 0$.

6. Find the equation of the plane passing through $(1, 2, 1)$ and containing the Y-axis. *Ans.* $x - z = 0$.

7. Show that the four planes $x + y + z - 3 = 0$, $2x + 2y - z + 1 = 0$, $x + y - 5z - 8 = 0$, and $3x + 3y + 7z - 4 = 0$ are concurrent.

CHAPTER XX

THE STRAIGHT LINE

113. Equations of a Line. Two intersecting planes determine a line L, and the general equations of the planes

$$(1) \qquad A_1x + B_1y + C_1z + D_1 = 0,$$
$$(2) \qquad A_2x + B_2y + C_2z + D_2 = 0$$

may be regarded as the general equations (simultaneous) of L. Since any two planes through L will determine L, the equations of a given line are not unique. Values of x, y, and z which satisfy both equations (1) and (2) are the coordinates of points on the line.

114. Special Forms of the Equations of a Line. Four special forms of the equations of a straight line in space are important.

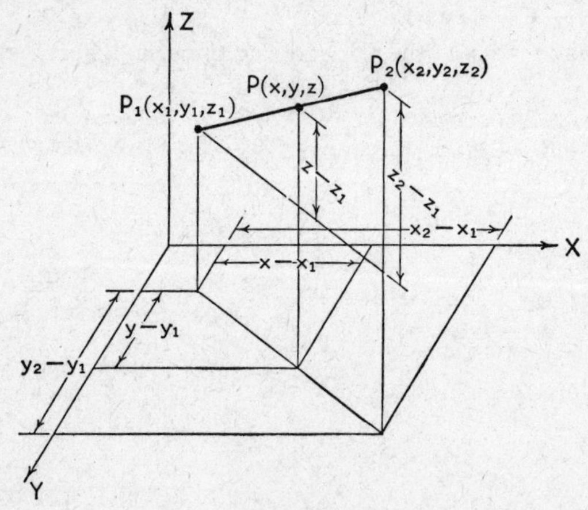

FIG. 185

I. *Two-point Form.* The equality of the ratios

$$(1) \qquad \frac{x - x_1}{x_2 - x_1} = \frac{y - y_1}{y_2 - y_1} = \frac{z - z_1}{z_2 - z_1}$$

is readily established from Fig. 185. These are the equations of the line joining the two points P_1 and P_2. The student should not be confused by the manner in which equations (1) are written; there are essentially only two equations and writing them in this particular form is shorter and avoids repetitions.

II. *Symmetric Form.* Since $x_2 - x_1 = a$, $y_2 - y_1 = b$, $z_2 - z_1 = c$ are direction numbers of the line we may write (1) in the symmetric form

$$\frac{x - x_1}{a} = \frac{y - y_1}{b} = \frac{z - z_1}{c}. \tag{2}$$

This is also called the point-direction form.

III. *Projection Form.* Note that

$$\frac{y - y_1}{y_2 - y_1} = \frac{z - z_1}{z_2 - z_1},$$
$$\frac{x - x_1}{x_2 - x_1} = \frac{z - z_1}{z_2 - z_1},$$
$$\frac{x - x_1}{x_2 - x_1} = \frac{y - y_1}{y_2 - y_1}$$

are respectively the equations of the planes through the line and perpendicular to the YZ-, XZ-, and XY-planes. These are of the form

$$By + Cz + D = 0,$$
$$Ax + Cz + D = 0,$$
$$Ax + By + D = 0,$$

and are called the *projecting planes* of the line, and any two of them are equations of the line.

IV. *Parametric Form.* If we set the ratios in (2) equal to a number, say t, we get

$$x = x_1 + at,$$
$$y = y_1 + bt,$$
$$z = z_1 + ct.$$

These are the parametric equations of a line with parameter t.

115. Reduction of General Form to Symmetric Form. Eliminating x from the general equations

$$A_1x + B_1y + C_1z + D_1 = 0, \tag{1}$$
$$A_2x + B_2y + C_2z + D_2 = 0, \tag{2}$$

we get

(3) $(A_1B_2 - A_2B_1)y + (A_1C_2 - A_2C_1)z + (A_1D_2 - A_2D_1) = 0.$

This is the YZ projecting plane; it and either (1) or (2) constitute equations of the line. By eliminating y from (1) and (2) we obtain

(4) $(B_1A_2 - B_2A_1)x + (B_1C_2 - B_2C_1)z + (B_1D_2 - B_2D_1) = 0,$

the projecting plane perpendicular to the XZ-plane. Equations (3) and (4) are equations of the line. Solving them for z gives equations of the form

$$z = \frac{x - x'}{a},$$

$$z = \frac{y - y'}{b}.$$

Hence the symmetric equations are

$$\frac{x - x'}{a} = \frac{y - y'}{b} = \frac{z}{1}. \tag{5}$$

To obtain (5) we began by eliminating x and y from the general equations. Now from (5) we see that $(x', y', 0)$ is the *piercing point* of the line in the XY-plane. Had we eliminated x and z we would have arrived at equations of the form

$$\frac{x - x''}{a} = \frac{y}{1} = \frac{z - z''}{c},$$

which reveal that the piercing point of the line in the XZ-plane is $(x'', 0, z'')$. Similarly for the piercing point in the YZ-plane.

Illustration 1. Write the equations of the line determined by $(-2, 3, 1)$ and $(4, -2, 5)$.

Solution. The two-point form yields

$$\frac{x + 2}{6} = \frac{y - 3}{-5} = \frac{z - 1}{4}.$$

Illustration 2. A line has direction numbers $1, -2, 6$ and passes through $(4, 5, -9)$. Find its equations and its direction cosines.

Solution. The equations of the line are

$$\frac{x - 4}{1} = \frac{y - 5}{-2} = \frac{z + 9}{6}.$$

The direction cosines are $\lambda = 1/\sqrt{41}$, $\mu = -2/\sqrt{41}$, $\nu = 6/\sqrt{41}$.

Illustration 3. Reduce the equations of the line $\pi_1 \equiv x - y - z + 3 = 0$, $\pi_2 \equiv 2x + y + 2z - 1 = 0$ to symmetric form.

Solution. Eliminating x we get

$$3y + 4z - 7 = 0,$$

$$z = \frac{y - \frac{7}{3}}{-\frac{4}{3}}. \tag{1}$$

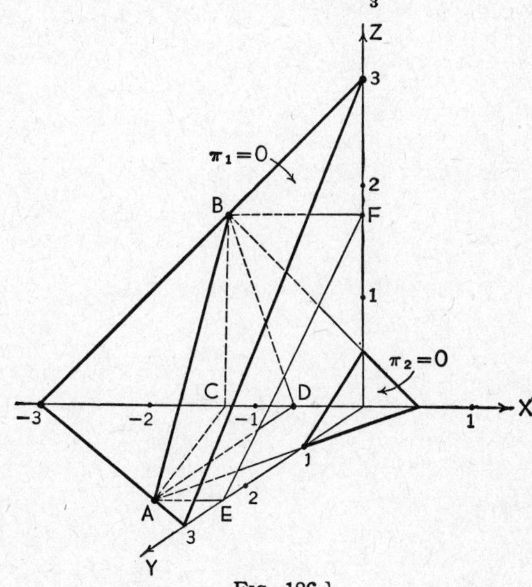

FIG. 186

Eliminating y we get

$$3x + z + 2 = 0,$$

$$z = \frac{x + \frac{2}{3}}{-\frac{1}{3}}. \tag{2}$$

Equations (1) and (2) are the projecting planes on the YZ- and XZ-planes respectively. Hence the symmetric equations of the line are

$$\frac{x + \frac{2}{3}}{-\frac{1}{3}} = \frac{y - \frac{7}{3}}{-\frac{4}{3}} = \frac{z}{1}.$$

The piercing point in the XY-plane is $(-\frac{2}{3}, \frac{7}{3}, 0)$. In Fig. 186 AB is the line, $ABFE$ is the YZ projecting plane, ABC is the XY projecting plane, ABD is the XZ projecting plane, A is the piercing point in the XY-plane.

Illustration 4. Write the equations of the line $x = 5 - 2t$, $y = 3 + t$, $z = -2 - 3t$ in symmetric form.

Solution. From $y = 3 + t$ we get $t = y - 3$. Substituting this in the other two parametric equations yields $x = 11 - 2y$, $z = 7 - 3y$ or

$$y = \frac{x - 11}{-2},$$

$$y = \frac{z - 7}{-3}.$$

Hence the symmetric equations are

$$\frac{x - 11}{2} = \frac{y}{-1} = \frac{z - 7}{3}.$$

The piercing point in the XZ-plane is (11, 0, 7).

EXERCISES

Write the equations of the following lines.

1. Through (7, − 1, 2), (3, 2, − 4). *Ans.* $\frac{x - 7}{4} = \frac{y + 1}{-3} = \frac{z - 2}{6}.$

2. Parallel to $\frac{x - 1}{2} = \frac{y + 4}{1} = \frac{z}{-3}$ through (4, 2, − 3).
Ans. $\frac{x - 4}{2} = \frac{y - 2}{1} = \frac{z + 3}{-3}.$

3. Through (1, 2, 8) and perpendicular to the plane $3x + 7y - 2z + 1 = 0$.
Ans. $\frac{x - 1}{3} = \frac{y - 2}{7} = \frac{z - 8}{-2}.$

4. Through (3, 4, 0) and perpendicular to the XY-plane.
Ans. $x = 3$, $y = 4$.

116. Angle between Line and Plane. Let the plane be $Ax + By + Cz + D = 0$ and the line $\frac{x - x_1}{a} = \frac{y - y_1}{b} = \frac{z - z_1}{c}.$ As in Illustration 2, § 108, we define the angle ϕ between a line and a plane to be the acute angle between the line and its projection upon the plane. Since this angle ϕ is the complement of the angle between the line and a normal to the plane it follows that

$$\sin \phi = \frac{|Aa + Bb + Cc|}{\sqrt{A^2 + B^2 + C^2}\sqrt{a^2 + b^2 + c^2}}.$$

117. Direction Numbers of the Line of Intersection of Two Planes. Let the planes be given in general form.

$$(1) \qquad A_1x + B_1y + C_1z + D_1 = 0,$$
$$(2) \qquad A_2x + B_2y + C_2z + D_2 = 0.$$

Of course (1) and (2) taken together are the general equations of the line L of intersection. Let direction numbers of the line be a, b, c. Now L is perpendicular to the normals to each plane. Therefore

$$(3) \qquad A_1a + B_1b + C_1c = 0,$$
$$(4) \qquad A_2a + B_2b + C_2c = 0.$$

Solving (3) and (4) for the ratios $a : b : c$ we get

$$(5) \qquad a : b : c = \begin{vmatrix} B_1 & C_1 \\ B_2 & C_2 \end{vmatrix} : - \begin{vmatrix} A_1 & C_1 \\ A_2 & C_2 \end{vmatrix} : \begin{vmatrix} A_1 & B_1 \\ A_2 & B_2 \end{vmatrix}.$$

118. Direction Numbers of the Normal to Two Skew Lines. There is one and only one line in space that is perpendicular to each of two skew lines and which at the same time intersects the skew lines. The segment thus cut off the common perpendicular is the distance between the lines.

The problem of determining direction numbers of the normal is essentially the same as that of determining direction numbers of the line of intersection of two planes, as was done in the preceding paragraph. For A_1, B_1, C_1 and A_2, B_2, C_2 are direction numbers of two certain lines and (5) of § 117 gives direction numbers a, b, c of the normal to these lines.

Illustration 1. Find the angle between the plane $2x-3y+z-4=0$ and the line $3x + y + 2z - 1 = 0$, $x + 7y - 2z + 5 = 0$.

Solution. Direction numbers of the normals to the plane are 2, −3, 1. Direction numbers of the line are

$$a : b : c = \begin{vmatrix} 1 & 2 \\ 7 & -2 \end{vmatrix} : - \begin{vmatrix} 3 & 2 \\ 1 & -2 \end{vmatrix} : \begin{vmatrix} 3 & 1 \\ 1 & 7 \end{vmatrix}.$$

Hence we may take

$$a = -16, \quad b = 8, \quad c = 20,$$

or

$$a = -4, \quad b = 2. \quad c = 5.$$

The angle ϕ is given by

$$\sin\phi = \frac{|-8-6+5|}{\sqrt{4+9+1}\sqrt{16+4+25}}$$
$$= \frac{9}{\sqrt{14}\sqrt{45}}$$
$$= \tfrac{3}{70}\sqrt{70}.$$

Illustration 2. Find the angle between the line joining $P_1(0, -5, 4)$ and $P_2(1, -4, 3)$ and the normal to the two lines I, $x + y + 5 = 0$, $3y+2z+7=0$ and II, $2x+2y+z-11=0$, $x+4y+z-16=0$.

Solution. Direction numbers are:

Of P_1P_2, 1, 1, -1;
Of I, 2, -2, 3;
Of II, 2, 1, -6;
Of the normal to I and II, $\begin{vmatrix} 1 & -6 \\ -2 & 3 \end{vmatrix}, -\begin{vmatrix} 2 & -6 \\ 2 & 3 \end{vmatrix}, \begin{vmatrix} 2 & 1 \\ 2 & -2 \end{vmatrix},$
or 3, 6, 2.

The angle between the two lines is given by

$$\cos\theta = \frac{3+6-2}{7\sqrt{3}}$$
$$= \frac{\sqrt{3}}{3}.$$

Illustration 3. Find the distance between the skew lines $x + y - 6 = 0$, $z - 2 = 0$ and $P_1(1, 2, 3)$, $P_2(-2, 7, 0)$.

Solution. A plane through the first line will be of the form

$$(x + y - 6) + k(z - 2) = 0.$$

This will be parallel to P_1P_2 if its normal is perpendicular to P_1P_2, that is, if

$$3 - 5 + 3k = 0,$$
$$k = \tfrac{2}{3}.$$

The plane through the first line and parallel to the second is therefore

$$3(x + y - 6) + 2(z - 2) = 0,$$
$$3x + 3y + 2z - 22 = 0.$$

The distance sought will be the distance from this plane to any point on P_1P_2. Taking P_1 itself we get for the distance

$$d = \frac{3+6+6-22}{\sqrt{22}}$$
$$= \frac{-7}{22}\sqrt{22}.$$

The negative sign indicates that the plane separates the origin and P_1. The numerical value of d is what was wanted.

EXERCISES

1. Find the angle between the XY-plane and the line $3x - y + 2z - 5 = 0$, $x + y - z + 6 = 0$. *Ans.* $\sin \phi = \frac{2}{21}\sqrt{42}$.

2. Find the angle between the two lines $x - y - 4z = 0$, $2x - y + 3z - 1 = 0$ and $x + y + z - 2 = 0$, $x - y + 5z - 5 = 0$. *Ans.* $\theta = 90°$.

3. Find the distance between the two lines $x - 2y + z + 9 = 0$, $2x + y - z - 10 = 0$ and $x + 3y + 2z + 5 = 0$, $3x + 4y + 2z + 1 = 0$.

Ans. 0.

4. Find the distance from the line $\frac{x-2}{1} = \frac{y+4}{-1} = \frac{z-6}{6}$ to the point $(-1, 1, 1)$.

Ans. $\sqrt{21}$.

5. Find the equations of the line which passes through $(-1, 2, -3)$ and which is parallel to each of the planes $6x - 3y + 2z - 8 = 0$ and $4x + y - 3z - 7 = 0$. *Ans.* $\frac{x+1}{7} = \frac{y-2}{26} = \frac{z+3}{18}$.

CHAPTER XXI

SPACE LOCI

119. Surfaces and Curves. In three dimensions we must consider two kinds of loci, namely surfaces and curves. The locus of a point which satisfies one condition $f(x, y, z) = 0$ is, in general, a surface. The locus of a point which satisfies simultaneously two conditions $f(x, y, z) = 0$, $g(x, y, z) = 0$ is, in general, a skew curve. An equation of the form $f + kg = 0$ is a surface passing through the curve $f = 0$, $g = 0$.

A skew curve may be represented parametrically by three equations of the form $x = f(t)$, $y = g(t)$, $z = h(t)$. (See parametric equations of a straight line, IV, § 114.) A surface may be represented parametrically by $x = f(t_1, t_2)$, $y = g(t_1, t_2)$, $z = h(t_1, t_2)$ where t_1 and t_2 are two independent parameters.

Rectangular, cylindrical, or spherical coordinates may be used in the representation of surfaces or curves.

120. Cylinders. A surface generated by a line moving parallel to a fixed line is called a *cylinder*. The moving line is called the *generator* (or generatrix) and it may be made to follow a given *directing curve* (or directrix), plane or skew. A particular generator, or position of it, is often referred to as an *element* of the cylinder.

Of special interest are the cylinders whose elements are perpendicular to a coordinate plane. An equation $f(x, y) = 0$, in the XY-plane, represents a curve. In three-space this same equation represents a cylinder since, for any point (x, y) on the curve in the XY-plane and regardless of the value of z, it is satisfied by the point (x, y, z). The cylinder is perpendicular to the XY-plane; its equation has no z-terms. This is typical and we build up the following outline.

Generator Perpendicular to	*Equation of Cylinder*
XY-plane	$f(x, y) = 0$, or $f(r, \theta) = 0$ (Cylindrical coordinates)
XZ-plane	$f(x, z) = 0$
YZ-plane	$f(y, z) = 0$

Illustration 1. Sketch the cylinder $y^2 = 4\,px$.

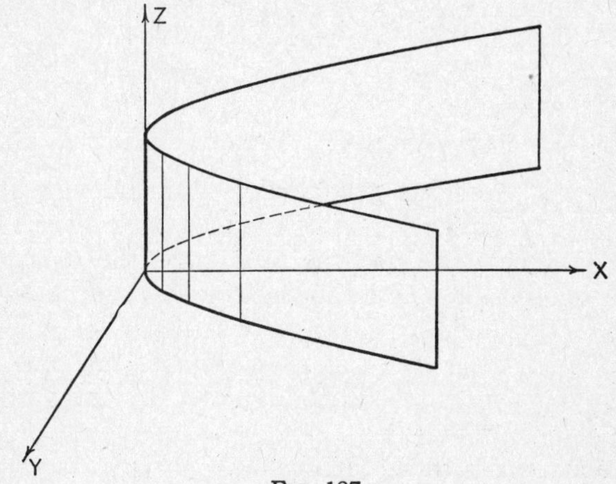

FIG. 187

Solution. In the XY-plane this is the equation of a parabola in standard form. Projecting this curve perpendicular to the XY-plane gives the parabolic cylinder shown in Fig. 187.

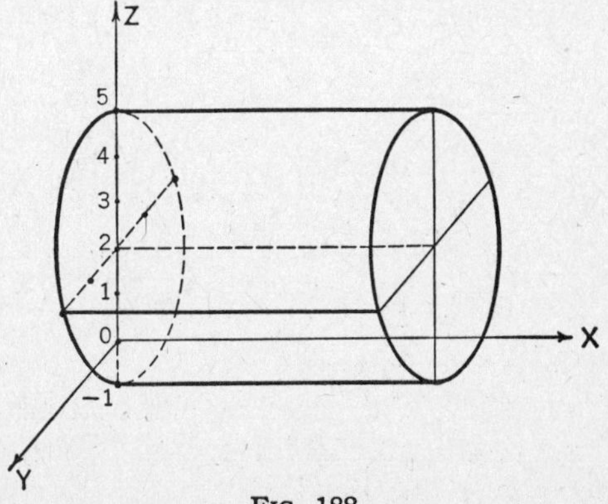

FIG. 188

Illustration 2. Write the equation of the cylinder perpendicular to the YZ-plane whose directing curve is the ellipse in the YZ-plane with major axis 6 in the z-direction and with minor axis 4, the center being at (0, 0, 2).

Solution. $$\frac{y^2}{4} + \frac{(z-2)^2}{9} = 1.$$

EXERCISES

Sketch the following cylinders.

1. $x^2 - z^2 = 1.$

2. $x^2 + y^2 = 1.$

3. $r = \cos 3\,\theta.$

121. Cones. A surface generated by a line moving about a fixed point on it is called a *cone*. The moving line is called the *generator* (or generatrix) and the fixed point the *vertex* of the cone. The generator may be made to follow a *directing curve* (or directrix), plane or skew. A particular generator, or position of it, is referred to as an *element* of the cone. The two similar portions of the cone separated by the vertex are called the *nappes* of the cone.

Every equation in three variables each of whose terms is of the second degree represents a cone with vertex at the origin. Consider as a special case the equation

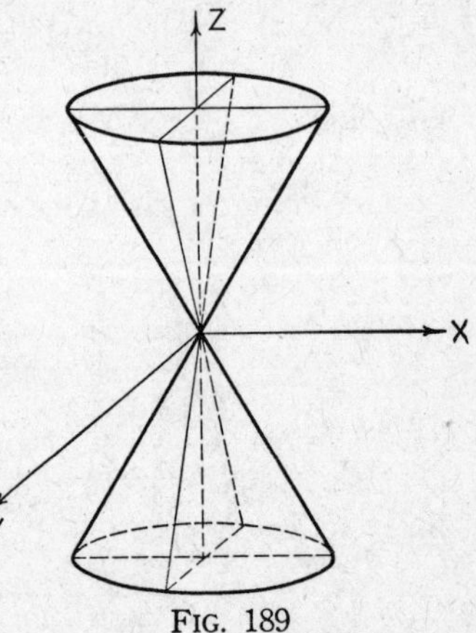

FIG. 189

$$\frac{x^2}{a^2} + \frac{y^2}{b^2} - \frac{z^2}{c^2} = 0. \tag{1}$$

If (x', y', z') is a point on this surface, so also is (kx', ky', kz'). Since every point of the line joining the origin and (x', y', z') has coordinates of the form (kx', ky', kz') the surface is a cone with its vertex at (0, 0, 0). It is called an elliptic cone since every cross section of it by a plane $z =$ const. is an ellipse. (See § 60.) The axis of the cone is the Z-axis.

Illustration. Write the equation of the cone with vertex at $V(-1, 1, 0)$ and with directing curve $4x^2 + z^2 - 4 = 0$, $y = 0$.

Solution. Let an element VP' connecting the vertex and any point $P'(x', y', z')$ on the directing curve pass through $P(x, y, z)$, a general point on the cone. The symmetric equations of this element are

$$\frac{x+1}{x'+1} = \frac{y-1}{y'-1} = \frac{z}{z'}. \tag{1}$$

Further, since P' is on the directing curve, we have

$$4x'^2 + z'^2 - 4 = 0, \quad y' = 0. \tag{2}$$

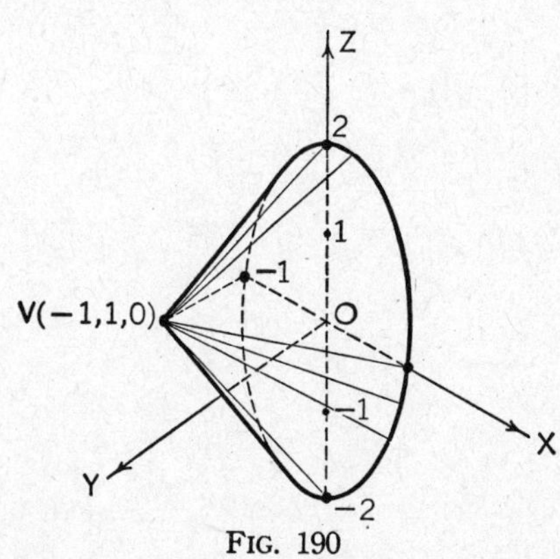

FIG. 190

We eliminate x', y', z' from the four conditions expressed in (1) and (2).

$$y' = 0,$$

$$z' = \frac{z}{(1-y)},$$

$$x' = \frac{(x+y)}{(1-y)},$$

$$4(x+y)^2 + z^2 - 4(1-y)^2 = 0.$$

This last is the equation of the cone.

EXERCISES

1. Sketch the cone $x^2 - y^2 + z^2 = 0$.

2. Find the equation of the cone with vertex at the origin and with directing curve $y^2 = z$, $x = 3$. *Ans.* $3y^2 = xz$.

122. Surfaces of Revolution. When a curve is revolved about a line a surface of revolution is generated. As a special case consider the curve $f(x, z) = 0$ in the XZ-plane and let this curve be revolved about the X-axis. A point $Q(x, z)$ on the curve will be turned into a typical position $P(x, y, Z)$ on the surface whose equation we seek. Now

$$AQ = AP$$

for all positions P. But

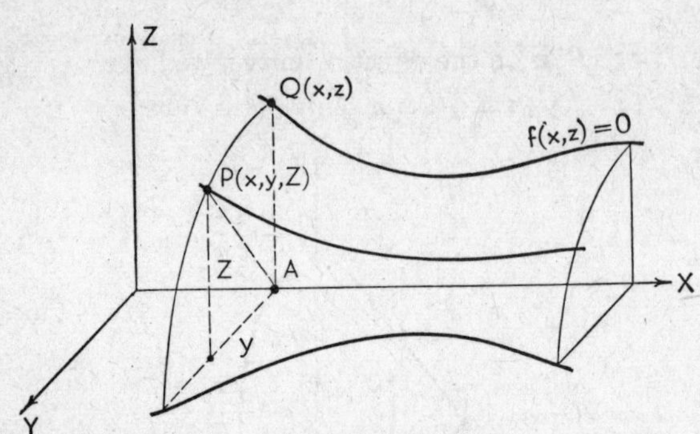

FIG. 191

(1) $$AP = \sqrt{y^2 + Z^2}$$

whereas

(2) $$AQ = z.$$

From (1) and (2) we see that in order to obtain the equation of the surface we must replace z in the equation of the curve by $\sqrt{y^2 + Z^2}$. We may, of course, then write z instead of Z. Hence if $f(x, z) = 0$ is the equation of a curve in the xz-plane, the equation of the surface obtained by revolving this curve about the X-axis is $f(x, \sqrt{y^2 + z^2}) = 0$.

In a similar manner we build up the following table.

Equation of Curve in	*Equation of Surface of Revolution about the*
XZ-plane, $f(x, z) = 0$ or XY-plane, $f(x, y) = 0$	X-axis, $f(x, \sqrt{y^2 + z^2}) = 0$
XZ-plane, $f(x, z) = 0$ or YZ-plane, $f(y, z) = 0$	Z-axis, $f(\sqrt{x^2 + y^2}, z) = 0$
XY-plane, $f(x, y) = 0$ or YZ-plane, $f(y, z) = 0$	Y-axis, $f(\sqrt{x^2 + z^2}, y) = 0$

Illustration 1. The line $y = mx$ is revolved about the X-axis. Find the equation of the right-circular cone thus generated.

Solution. We substitute in the equation $y = mx$ the quantity $\sqrt{y^2 + z^2}$ for y.

$$\sqrt{y^2 + z^2} = mx,$$
$$m^2x^2 - y^2 - z^2 = 0.$$

Illustration 2. The hypocycloid $y^{\frac{2}{3}} + z^{\frac{2}{3}} = a^{\frac{2}{3}}$ is revolved about the Z-axis. Find the equation of the surface generated.

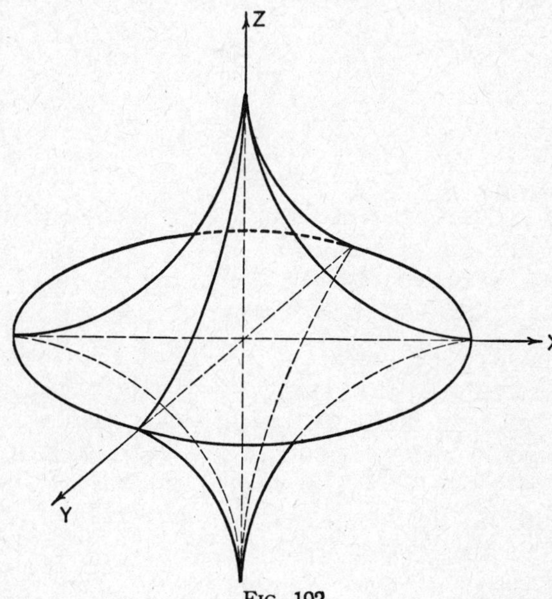

FIG. 192

Solution.

$$[(x^2 + y^2)^{\frac{1}{2}}]^{\frac{2}{3}} + z^{\frac{2}{3}} = a^{\frac{2}{3}},$$
$$(x^2 + y^2)^{\frac{1}{3}} + z^{\frac{2}{3}} = a^{\frac{2}{3}}.$$

Illustration 3. The circle $(x - a)^2 + z^2 = b^2$ is revolved around the Z-axis. Find the equation of the *torus* thus generated.

Solution. The surface looks like a doughnut or an inner tube and has the equation

$$(\sqrt{x^2 + y^2} - a)^2 + z^2 = b^2,$$
$$x^2 + y^2 + a^2 - 2\,a\sqrt{x^2 + y^2} = b^2 - z^2,$$
$$(x^2 + y^2 + z^2 + a^2 - b^2)^2 = 4\,a^2(x^2 + y^2).$$

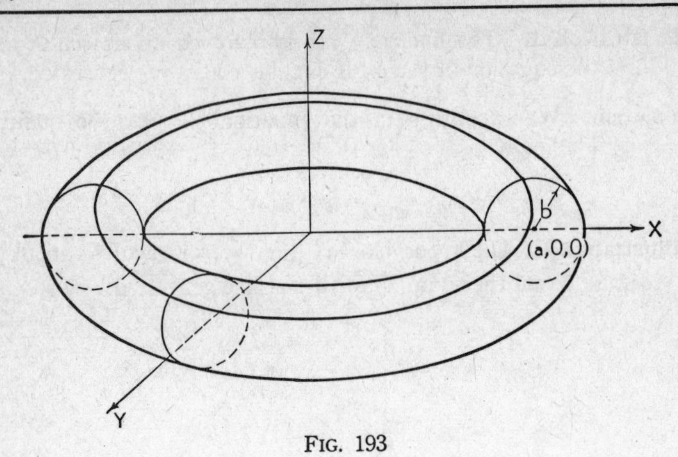

FIG. 193

EXERCISES

Find the equations of the following surfaces of revolution and sketch.

1. $z^2 = y,\ x = 0$ about the Z-axis. *Ans.* $z^4 = x^2 + y^2$.

2. $z^2 = y,\ x = 0$ about the Y-axis. *Ans.* $x^2 + z^2 = y$.

3. $x^2 + y^2 = 1,\ z = 0$ about the X-axis. *Ans.* $x^2 + y^2 + z^2 = 1$.

123. Sketching a Surface. Although the pictorial representation of three dimensions in two and the sketching of surfaces are more difficult than plotting curves in a plane, yet the procedures are essentially the same. These include the determination of: I, *intercepts and traces;* II, *extent;* III, *symmetry;* IV, *asymptotes;* V, *plane sections.* Let the equation of the surface be $F(x, y, z) = 0$. We summarize the analyses as follows:

I. *Intercepts and Traces.* For the intercepts on the X-axis set $y = 0$, $z = 0$ and solve $F(x, 0, 0) = 0$ for the values of x. Similarly for the other intercepts.

The traces of a surface are the curves of intersection with the coordinate planes. For the xy-trace set $z = 0$ and sketch the curve $F(x, y, 0) = 0$ *in* the XY-plane. Similarly for the other traces.

II. *Extent.* As in plotting in two dimensions determine whether the surface is bounded. This is readily done in connection with V.

III. *Symmetry.* The tests are as follows:

Symmetry with respect to	*If* $F(x, y, z) \equiv$
YZ-plane	$F(-x, y, z)$
XZ-plane	$F(x, -y, z)$
XY-plane	$F(x, y, -z)$
Z-axis	$F(-x, -y, z)$
Y-axis	$F(-x, y, -z)$
X-axis	$F(x, -y, -z)$
Origin	$F(-x, -y, -z)$

IV. *Asymptotes.* In only a few instances will we treat surfaces with asymptotes. Cylinders $F(x, y) = 0$ whose traces in the XY-plane have asymptotic lines will have corresponding asymptotic planes. Similarly for cylinders perpendicular to the other coordinate planes. In § 129 we shall investigate a surface that has an asymptotic cone.

V. *Plane Sections.* The most fruitful method of studying a surface is by means of its plane sections. In general these

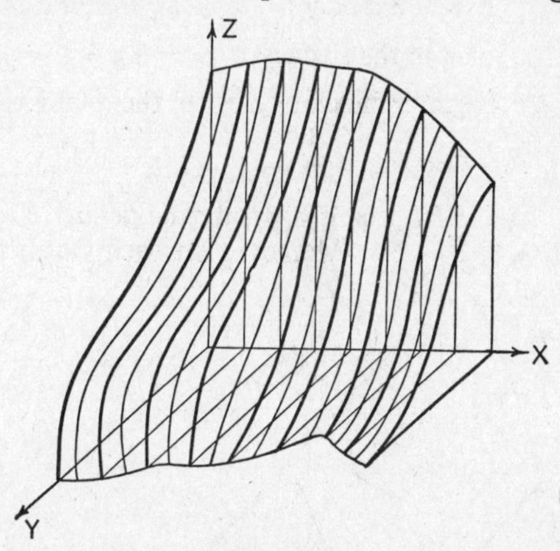

Fig. 194

sections are by planes parallel to the coordinate planes, each section yielding a plane curve. We thus reduce the problem of plotting a surface to that of plotting parallel sections (curves) of it. If the cutting planes are taken sufficiently close the curves in them will indicate the shape of the surface just as isolated but sufficiently close points in a plane will indicate the shape of a curve.

To study the sections parallel to the yz-plane we set $x = k$ and sketch the curves $F(k, y, z) = 0$ for different values of k.

Illustration 1. Sketch the surface $y(x^2 - 4) - (x + 1)(x - 3) = 0$.

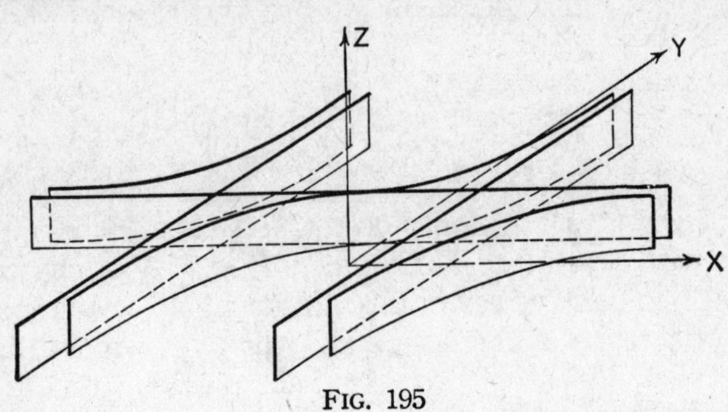

FIG. 195

Solution. This is a cylinder perpendicular to the XY-plane since z is missing. (The xy-trace was plotted in Fig. 31, p. 37.) The planes $x = -2$, $x = 2$, and $y = 1$ are asymptotic planes.

Illustration 2. Sketch the surface $x + y^2 + z - 1 = 0$.

Solution. We begin immediately with the plane sections. For $x = k$ we get

$$y^2 = -[z - (1 - k)].$$

For a given value of k this is a parabola, opening downward, with vertex at $(k, 0, 1 - k)$. Moreover for each parabola $p = -\frac{1}{4}$ so

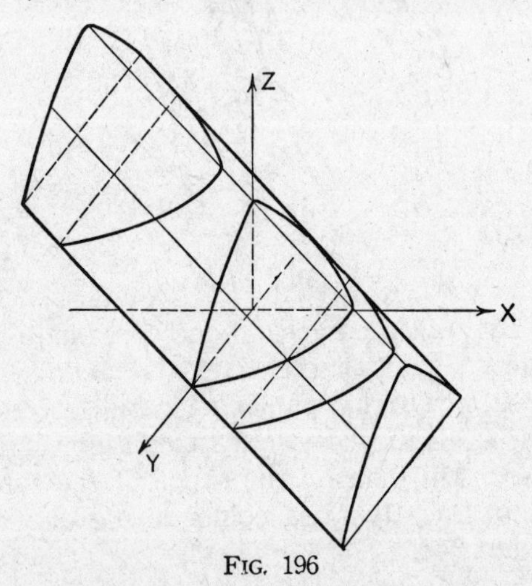

FIG. 196

that the parabolas are all congruent. This information alone is enough to make a reasonable graph of the surface.

For $y = k$, we have

$$x + z = 1 - k^2,$$

which is a set of straight lines. Any two of them are parallel.

The planes $z = k$ cut out the parabolas

$$y^2 = -[x - (1 - k)].$$

There is symmetry with respect to the XZ-plane. The surface is a parabolic cylinder with generators $x + z = 1 - k^2$, $y = k$ and directing curve $y^2 = -(z - 1)$, $x = 0$.

Illustration 3. Sketch the surface $x^2 + y^2 = z$.

Solution. The surface passes through the origin and lies wholly above the XY-plane since z must be positive or zero. There is symmetry with respect to the XZ- and YZ-planes and the Z-axis. Indeed the surface is one of revolution about the Z-axis and we may think of $x^2 = z$ as the generating curve in the XZ-plane.

Sections for $x = k$ are the parabolas.

$$y^2 = (z - k^2).$$

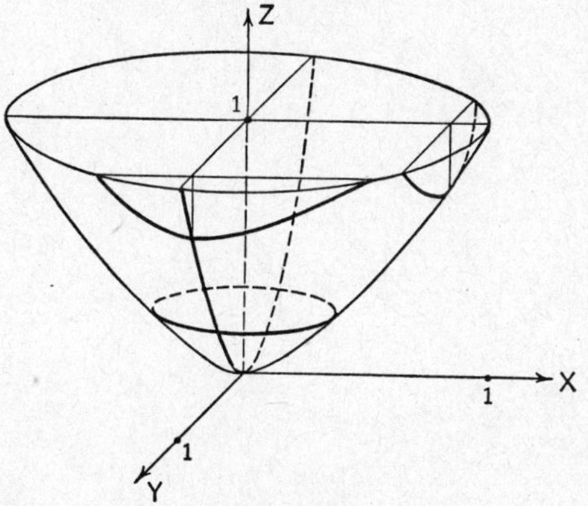

FIG. 197

Sections for $y = k$ are the parabolas

$$x^2 = (z - k^2).$$

Sections for $z = k$ are the circles

$$x^2 + y^2 = k.$$

The surface is a circular paraboloid.

EXERCISES

Sketch the following surfaces.

1. $4x^2 + y^2 = 4$.

2. $x^2 + y^2 + z^2 = 1$.

3. $xy = 1$.

4. $(x + y)^2 = 1 - z$.

124. Sketching a Curve. A curve is represented analytically by the equations of two intersecting surfaces $f = 0$, $g = 0$. In general, surfaces intersect in skew (twisted) curves; in special cases the curves may lie in a plane. Since any two surfaces intersecting in a given curve will define the curve we may find it useful, for purposes of plotting, to replace the given surfaces

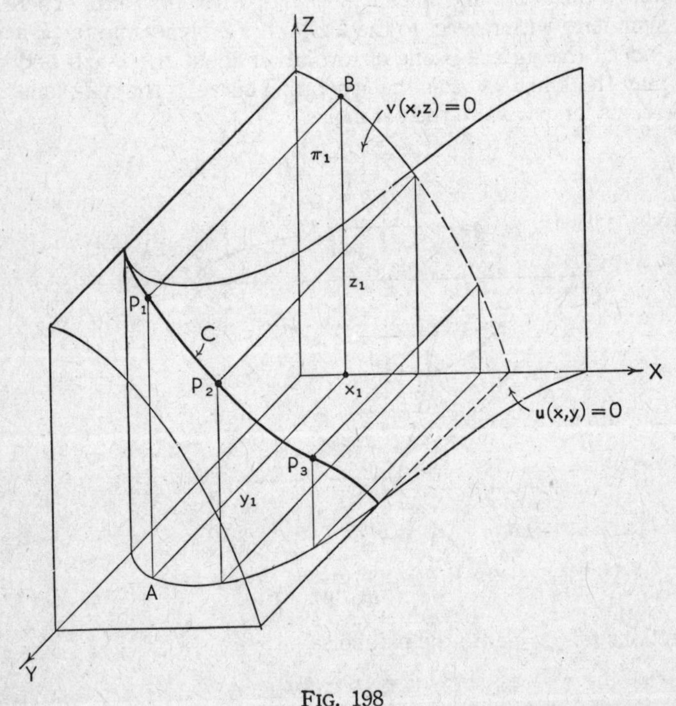

FIG. 198

by others. By eliminating x between $f = 0$ and $g = 0$ we obtain a cylinder perpendicular to the YZ-plane and passing through the curve. This is called the *projecting cylinder* of the curve, which may be considered the directrix of the cylinder. Similarly we may obtain the other projecting cylinders. Two of the three projecting cylinders may be used to advantage in plotting the curve C as follows. In Fig. 198 let portions of the projecting cylinders in the first octant perpendicular to the xy- and xz-planes be

$$f + k_1 g \equiv u(x, y) = 0,$$
$$f + k_2 g \equiv v(x, z) = 0$$

respectively. Pass a plane π_1 perpendicular to the x-axis through $(x_1, 0, 0)$ cutting out on the projecting cylinders the generators AP_1 and BP_1. P_1 is a point on the curve C of intersection of $u = 0$, $v = 0$ and hence of the original surfaces $f = 0$, $g = 0$. In this manner as many points P_1, P_2, P_3, $\cdots$ may be constructed as are needed to insure a smooth graph. It is desirable to retain, if possible, the graphs of the original surfaces.

Since to sketch a projecting cylinder is essentially to draw a plane curve, the problem of plotting a twisted curve in space is reduced to that of plotting plane curves. The attendant difficulties are only slightly increased by the necessity of perspective drawings.

Illustration 1. Sketch the curve (1), $x^2 + y^2 + z^2 = 4$, (2), $(x - 1)^2 + y^2 = 1$.

Solution. The first equation evidently represents a sphere of radius 2 with center at the origin since the distance from the origin to any point (x, y, z) on it is $\sqrt{x^2 + y^2 + z^2}$ and this, being equal to 2, yields equation (1). The second surface is that of a cylinder perpendicular to the XY-plane whose trace in that plane is the circle of radius 1 with center at $(1, 0)$. Hence (2) is already the xy projecting cylinder of the curve of intersection. By eliminating y between (1) and (2) we get the equation of the xz projecting cylinder. Thus

$$x^2 + [1 - (x - 1)^2] + z^2 = 4$$
$$z^2 = -2(x - 2).$$

The xz projecting cylinder is parabolic therefore with xz-trace AA' in Fig. 199. The location of points P_1, P_1', etc. follows the method outlined above.

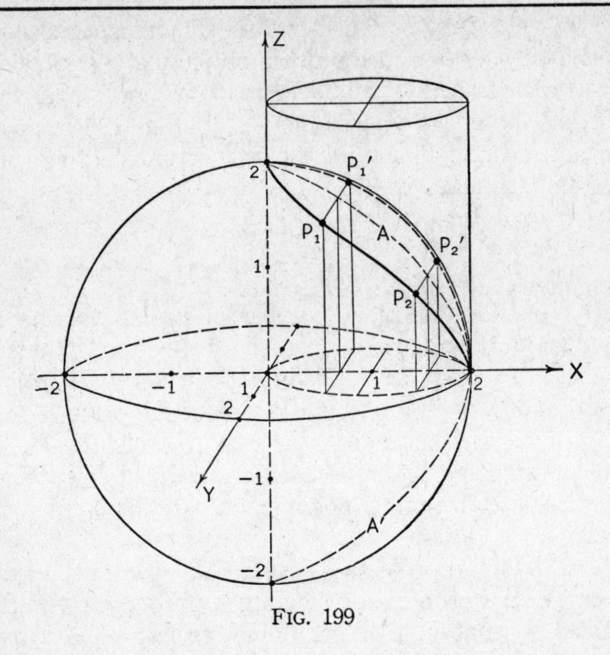

FIG. 199

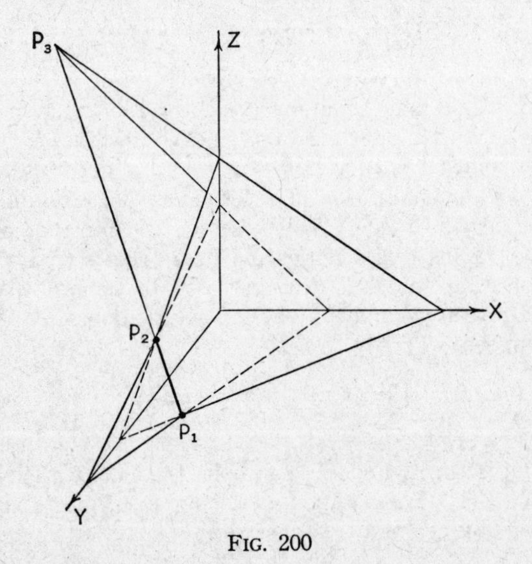

FIG. 200

Illustration 2. Draw the line of intersection of two planes.

Solution. This elementary problem requires no elaborate theory of projecting planes although the method would apply. Write the equations of the planes in intercept form

$$\frac{x}{a_1} + \frac{y}{b_1} + \frac{z}{c_1} = 1,$$

$$\frac{x}{a_2} + \frac{y}{b_2} + \frac{z}{c_2} = 1.$$

Sketch each plane; where the traces intersect we have the piercing points P_1, P_2, P_3 and the join of them is the line of intersection.

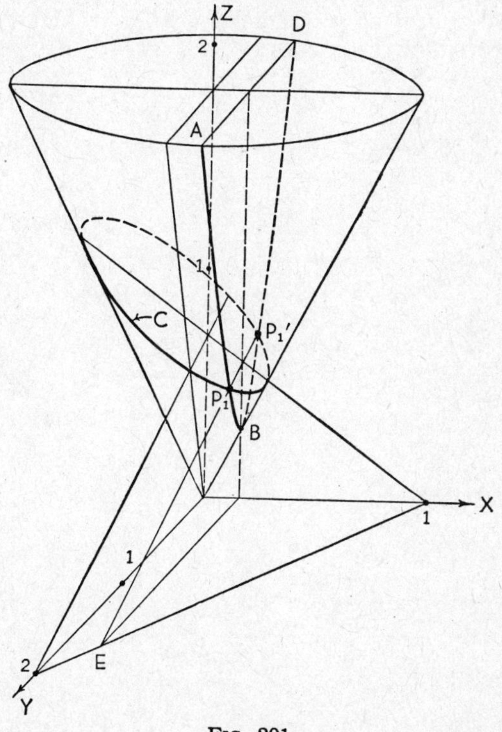

FIG. 201

Illustration 3. Find the curve C of intersection of the cone (1), $4x^2 + 4y^2 - z^2 = 0$ and the plane (2), $6x + 3y + 8z - 6 = 0$.

Solution. The method of projecting cylinders is numerically involved. For example, the equation of the xy-trace of the xy projecting cylinder is

$$220x^2 - 36xy + 247y^2 + 72x + 36y - 36 = 0.$$

This is an ellipse but it is troublesome to sketch without a rotation of axes.

Instead we proceed as follows. Cut the cone with the plane $x = k$. The equation of the curve cut out is

$$4y^2 - z^2 = -4k^2,$$

which is a hyperbola ABD. (See § 60.) The plane $x = k$ cuts the plane (2) in the straight line P_1E and the intersections of this line and the hyperbola yield points P_1 and P_1' on the curve sought. Other points on C can be constructed in like manner.

EXERCISES

Sketch the following curves.

1. $x^2 + z^2 = 1,\ x^2 + y^2 = 1.$

2. $x^2 + y^2 = z,\ x + y + z = 1.$

3. $x^2 + y^2 + z^2 = 1,\ x^2 + y^2 = z.$

4. The locus of a point which moves on the cylinder $x^2 + y^2 = a^2$ so that the distance it moves parallel to the axis of the cylinder is directly proportional to the angle through which it rotates about the axis is called a *circular helix*. Derive the parametric equations of this skew curve.

Ans. $x = a\cos\theta,\ y = a\sin\theta,\ z = k\theta.$

CHAPTER XXII

THE QUADRIC SURFACES

125. General Equation of the Second Degree. We have seen that the general linear equation represents a plane and that two such determine a line. The general equation of the second degree is

$$Ax^2 + By^2 + Cz^2 + Dxy + Eyz + Fxz + Gx + Hy + Kz + L = 0;$$

its graph is called the *quadric surface.* This surface is such that any plane section of it is a conic, for which reason it is also called a *conicoid.* A few examples were met in the preceding chapter; we now discuss each of the quadrics. Since the methods of analysis have been fully illustrated in Chapter XXI only the minimum outlines will be given here.

126. The Ellipsoid. Consider the special equation

$$\frac{x^2}{a^2} + \frac{y^2}{b^2} + \frac{z^2}{c^2} = 1. \tag{1}$$

The section cut out by the plane $z = 0$ is the ellipse $x^2/a^2 + y^2/b^2 = 1$ with semiaxes a and b. The section by the plane $z = k$ is the ellipse

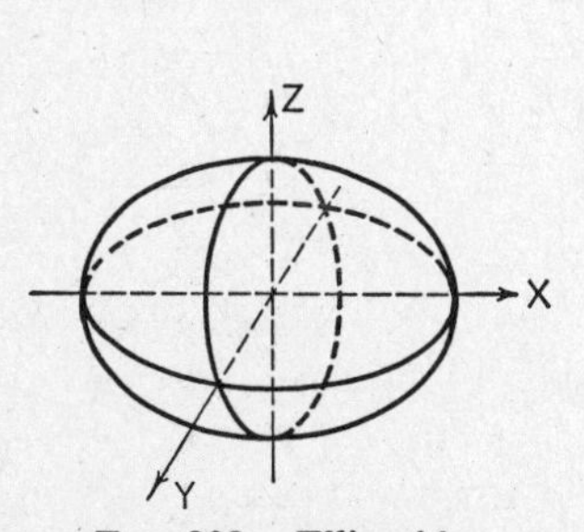

FIG. 202. Ellipsoid: $\frac{x^2}{a^2} + \frac{y^2}{b^2} + \frac{z^2}{c^2} = 1.$

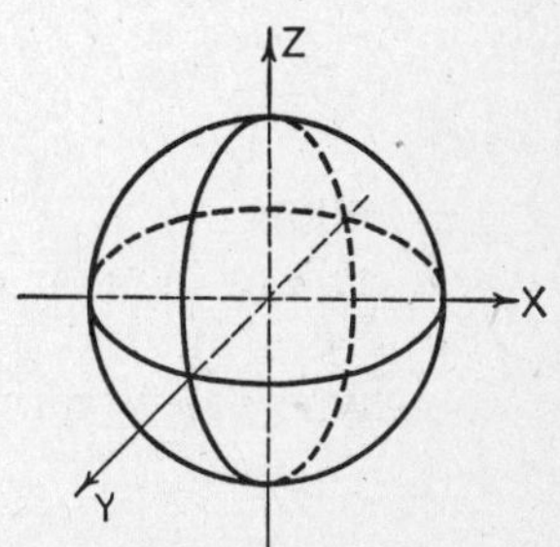

FIG. 203. Sphere: $x^2 + y^2 + z^2 = a^2.$

(2)
$$\frac{x^2}{a^2} + \frac{y^2}{b^2} = 1 - \frac{k^2}{c^2},$$

or

(3)
$$\frac{x^2}{\frac{a^2}{c^2}(c^2 - k^2)} + \frac{y^2}{\frac{b^2}{c^2}(c^2 - k^2)} = 1$$

with semiaxes $\frac{a}{c}\sqrt{c^2 - k^2}$, $\frac{b}{c}\sqrt{c^2 - k^2}$. For $k = c$, equation (2) shows that $x = 0$, $y = 0$. The surface is bounded above and below since $|z|$ must not exceed c.

Similarly the sections cut out by the planes $y = k$ and $x = k$ are ellipses and $|y| \leqq b$, $|x| \leqq a$. The surface is symmetric with respect to the coordinate planes, the axes, and the origin. The semiaxes of the ellipsoid are a, b, c and the center is at the origin.

For $a = b \neq c$ the surface is one of revolution and is called a spheroid (oblate if $a > c$, prolate if $a < c$). For $a = b = c$ the surface is a sphere. The equation of the sphere in spherical coordinates, with center at the origin and radius a, is $\rho = a$.

The equation

(4)
$$\frac{(x - h)^2}{a^2} + \frac{(y - k)^2}{b^2} + \frac{(z - l)^2}{c^2} = 1$$

represents the same ellipsoid with the same orientation of axes but with center at (h, k, l).

127. The Hyperboloid of One Sheet. The sections of

(1)
$$\frac{x^2}{a^2} + \frac{y^2}{b^2} - \frac{z^2}{c^2} = 1$$

are, for

$z = k$, an ellipse,
$y = k$, a hyperbola (pair of straight lines if $k = b$),
$x = k$, a hyperbola (pair of straight lines if $k = a$).

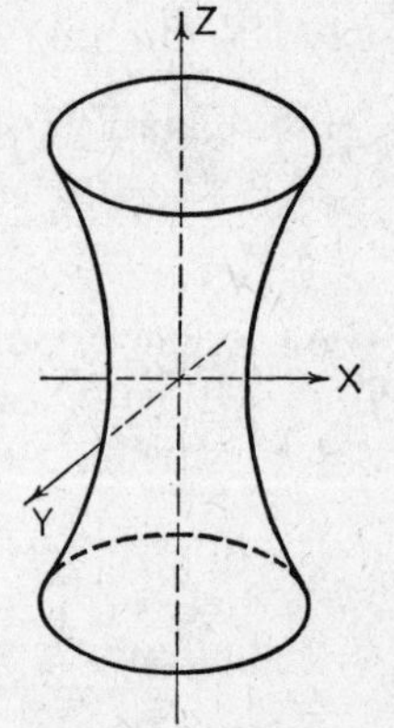

FIG. 204. Hyperboloid of One Sheet: $\frac{x^2}{a^2} + \frac{y^2}{b^2} - \frac{z^2}{c^2} = 1.$

Since the sections two ways are hyperbolas, the surface is called a hyperboloid. The surface is connected (one sheet) but is unbounded. The smallest elliptical section occurs for $z = 0$, the xy-trace. There is symmetry with respect to the coordinate planes, the axes, and the origin. The axis of the hyperboloid corresponds to the term with the minus sign and the center is at the origin.

The equation

$$\frac{(x-h)^2}{a^2} + \frac{(y-k)^2}{b^2} - \frac{(z-l)^2}{c^2} = 1 \tag{2}$$

represents the same hyperboloid with the same orientation but with center at (h, k, l).

128. The Hyperboloid of Two Sheets. The sections of

$$\frac{x^2}{a^2} - \frac{y^2}{b^2} - \frac{z^2}{c^2} = 1 \tag{1}$$

are, for

$z = k$, a hyperbola (pair of straight lines if $k = c$),
$y = k$, a hyperbola (pair of straight lines if $k = b$),
$x = k$, an ellipse.

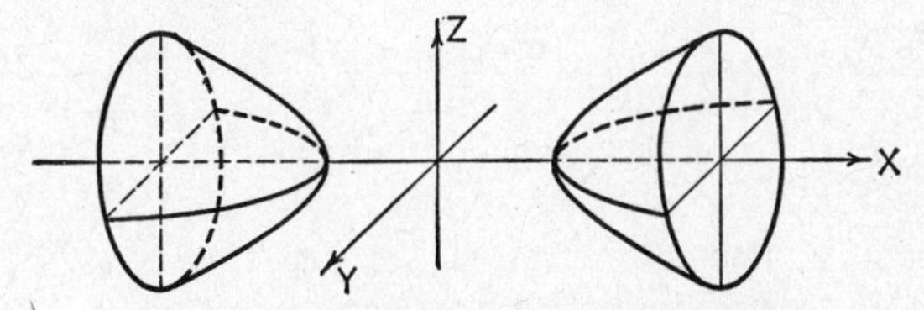

FIG. 205. Hyperboloid of Two Sheets:
$$\frac{x^2}{a^2} - \frac{y^2}{b^2} - \frac{z^2}{c^2} = 1.$$

This, too, is a hyperboloid; but since for sections $x = k$, $|k|$ cannot be less than a, the surface is made up of two pieces or sheets. Again there is symmetry with respect to the coordinate planes, the axes, and the origin. The axis corresponds to the term with the plus sign, the center is at the origin, and the points $(\pm a, 0, 0)$ are called vertices.

The equation

$$\frac{(x-h)^2}{a^2} - \frac{(y-k)^2}{b^2} - \frac{(z-l)^2}{c^2} = 1 \tag{2}$$

represents the same hyperboloid with the same orientation but with center at (h, k, l).

129. The Cone. The sections of

$$\frac{x^2}{a^2} + \frac{y^2}{b^2} - \frac{z^2}{c^2} = 0 \tag{1}$$

are, for

$z = k$, an ellipse,
$y = k$, a hyperbola (pair of straight lines if $k = 0$),
$x = k$, a hyperbola (pair of straight lines if $k = 0$).

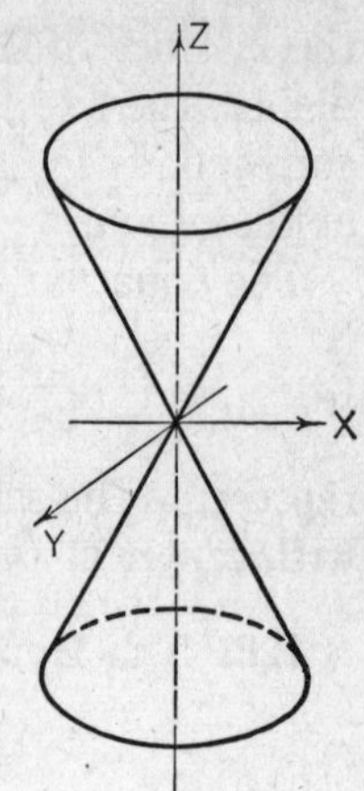

Fig. 206. Elliptic Cone. $\frac{x^2}{a^2} + \frac{y^2}{b^2} - \frac{z^2}{c^2} = 0.$

In § 121 we showed this to be a cone with axis coinciding with the Z-axis and with vertex at the origin. It is a degenerate quadric surface and is a special case of a hyperboloid, as can be seen by considering the limiting case of the equation

$$\frac{x^2}{a^2} + \frac{y^2}{b^2} - \frac{z^2}{c^2} = d \tag{2}$$

as $d \to 0$. For $d > 0$, (2) represents a hyperboloid of one sheet with axis coinciding with the Z-axis; for $d < 0$, (2) is a hyperboloid of two sheets with axis coinciding with the Z-axis. For $d = 0$, (2) is a cone asymptotic to each hyperboloid. The plane analogue to the cone is the pair of asymptotes to conjugate hyperbolas. (See § 55.)

The equation

$$\frac{(x-h)^2}{a^2} + \frac{(y-k)^2}{b^2} - \frac{(z-l)^2}{c^2} = 0 \tag{3}$$

represents the same cone with the same orientation but with the vertex at (h, k, l).

130. The Elliptic Paraboloid. The sections of

$$\frac{x^2}{a^2} + \frac{y^2}{b^2} = cz \tag{1}$$

are, for

$z = k$, an ellipse,
$y = k$, a parabola,
$x = k$, a parabola.

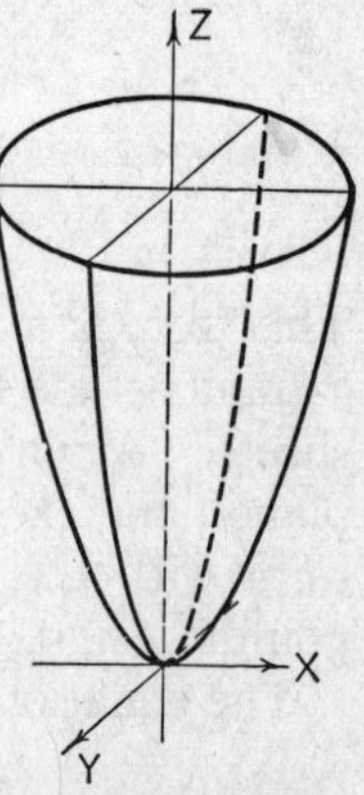

Fig. 207. Elliptic Paraboloid: $\frac{x^2}{a^2} + \frac{y^2}{b^2} = cz.$

The surface is called an elliptic paraboloid. Its axis coincides with the axis of the term of 1st degree, it lies wholly above the XY-plane (for $c < 0$), and its vertex is at the origin.

The equation

$$\frac{(x-h)^2}{a^2} + \frac{(y-k)^2}{b^2} = c(z-l) \tag{2}$$

represents the same paraboloid with the same orientation but with vertex at (h, k, l).

131. The Hyperbolic Paraboloid. The sections of

$$\frac{x^2}{a^2} - \frac{y^2}{b^2} = cz \tag{1}$$

are, for

$z = k$, a hyperbola (pair of straight lines if $k = 0$),
$y = k$, a parabola,
$x = k$, a parabola.

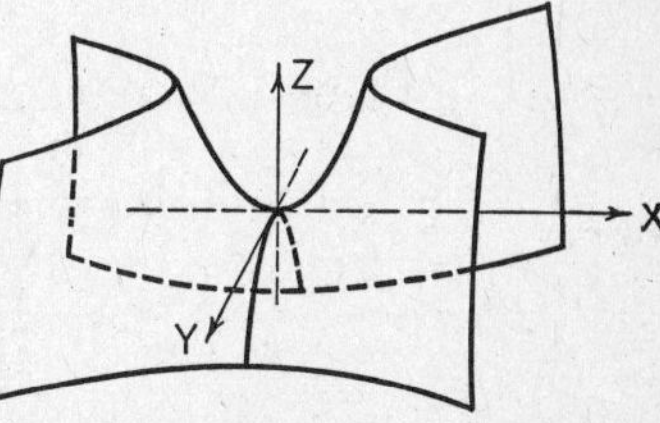

FIG. 208. Hyperbolic Paraboloid: $\frac{x^2}{a^2} - \frac{y^2}{b^2} = cz.$

This surface is called a hyperbolic paraboloid. It is saddle-shaped and the point on it at the origin is called a saddle point.

The equation

$$\frac{(x-h)^2}{a^2} - \frac{(y-k)^2}{b^2} = c(z-l) \tag{2}$$

represents the same surface with the same orientation but with saddle point at (h, k, l).

132. The Cylinders. We have already discussed the general cylinders in § 120. The quadric cylinders (degenerate quadric surfaces) are

$\frac{x^2}{a^2} + \frac{y^2}{b^2} = 1$, elliptic cylinder;

$\frac{x^2}{a^2} - \frac{y^2}{b^2} = 1$, hyperbolic cylinder;

$y^2 = 4px$, parabolic cylinder;

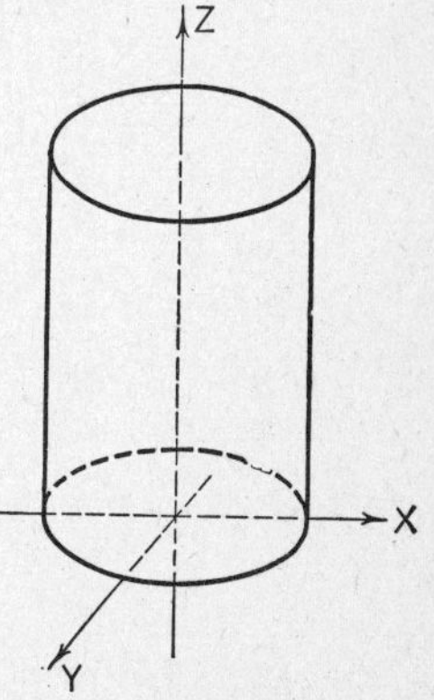

FIG. 209. Elliptic Cylinder: $\frac{x^2}{a^2} + \frac{y^2}{b^2} = 1.$

$\frac{(x-h)^2}{a^2} + \frac{(y-k)^2}{b^2} = 1,$ elliptic cylinder with axis $x = h$, $y = k$;

$\frac{(x-h)^2}{a^2} - \frac{(y-k)^2}{b^2} = 1,$ hyperbolic cylinder with axis $x = h$, $y = k$;

$(y-k)^2 = 4p(x-h),$ parabolic cylinder with line of vertices $x = h$, $y = k$.

133. Résumé. We give a complete list of forms to which a second-degree equation can be reduced (by rotations and translations). The simplest standard forms of the equations are used.

NON-DEGENERATE QUADRIC SURFACES

Central Quadrics	$\frac{x^2}{a^2} + \frac{y^2}{b^2} + \frac{z^2}{c^2} = \pm 1,$	Ellipsoid (real or imaginary)
	$\frac{x^2}{a^2} + \frac{y^2}{b^2} - \frac{z^2}{c^2} = \pm 1,$	Hyperboloid (one and two sheets respectively)
Non-Central Quadrics	$\frac{x^2}{a^2} + \frac{y^2}{b^2} = cz,$	Elliptic Paraboloid
	$\frac{x^2}{a^2} - \frac{y^2}{b^2} = cz,$	Hyperbolic Paraboloid

DEGENERATE QUADRIC SURFACES

$\frac{x^2}{a^2} + \frac{y^2}{b^2} \mp \frac{z^2}{c^2} = 0,$	Cone (real or imaginary)
$\frac{x^2}{a^2} + \frac{y^2}{b^2} = \pm 1,$	Elliptic Cylinder (real or imaginary)
$\frac{x^2}{a^2} - \frac{y^2}{b^2} = 1,$	Hyperbolic Cylinder
$\frac{x^2}{a^2} \mp \frac{y^2}{b^2} = 0,$	Intersecting Planes (real or imaginary)
$y^2 = 4px,$	Parabolic Cylinder
$y^2 = a,$	Parallel Planes (real or imaginary)
$y^2 = 0,$	Coinciding Planes

Illustration 1. Identify and sketch $x^2 - y^2 + 2z^2 - 4x - 6y - 8z - 3 = 0$.

Solution. We complete the squares on the x's, y's, and z's separately, obtaining

$$(x-2)^2 - (y+3)^2 + 2(z-2)^2 = 3 + 4 - 9 + 8,$$

or

$$\frac{(x-2)^2}{6} - \frac{(y+3)^2}{6} + \frac{(z-2)^2}{3} = 1.$$

Hence the surface is a hyperboloid of one sheet with center at $(2, -3, 2)$ and axis perpendicular to the XZ-plane. It is left as an exercise for the student to sketch the graph.

Illustration 2. Identify and sketch $x^2 + z^2 - 3x - y + z - 1 = 0$.

Solution. Completing the squares we get

$$(x - \tfrac{3}{2})^2 + (z + \tfrac{1}{2})^2 = y + \tfrac{7}{2}.$$

This represents an elliptic paraboloid with vertex at $(\frac{3}{2}, -\frac{7}{2}, -\frac{1}{2})$ opening up in the direction of the $+Y$-axis.

Illustration 3. Identify and sketch $4x^2 - 3y^2 + 12z^2 - 16x - 30y - 24z - 47 = 0$.

Solution. Completing the squares we get

$$4(x-2)^2 - 3(y+5)^2 + 12(z-1)^2 = 47 + 16 - 75 + 12 = 0,$$

or $$\frac{(x-2)^2}{3} - \frac{(y+5)^2}{4} + \frac{(z-1)^2}{1} = 0.$$

This is a cone with vertex at $(2, -5, 1)$ and axis perpendicular to the XZ-plane.

EXERCISES

Identify and sketch.

1. $4y^2 - z^2 - 12x - 8y - 4z = 0$.

Ans. The hyperbolic paraboloid $\frac{(y-1)^2}{1} - \frac{(z+2)^2}{4} = 3x$.

2. $6x^2 + 3y^2 + 2z^2 - 6y - 3 = 0$.

Ans. The ellipsoid $\frac{x^2}{1} + \frac{(y-1)^2}{2} + \frac{z^2}{3} = 1$.

3. $z^2 - x^2 = 0$. *Ans.* A pair of planes.

4. $2x^2 - y^2 - 2z^2 - 4y - 6 = 0$.

Ans. The hyperboloid of two sheets $\frac{x^2}{1} - \frac{(y+2)^2}{2} - \frac{z^2}{1} = 1$.

5. Write the equation of the sphere with center at $(2, -5, 1)$ and radius 6.

Ans. $(x-2)^2 + (y+5)^2 + (z-1)^2 = 36$.

134. Ruled Surfaces. A surface such that through any point on it there passes a line lying wholly on the surface is called a *ruled surface.* Cones and cylinders are ruled surfaces and so are the hyperboloid of one sheet and the hyperbolic paraboloid.

Equation (1), § 127, of the hyperboloid of one sheet can be written in the form

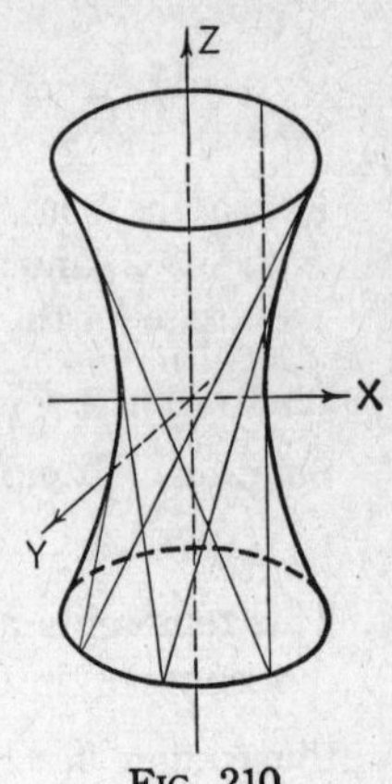

FIG. 210

$$(1)\quad \left(\frac{x}{a}+\frac{z}{c}\right)\left(\frac{x}{a}-\frac{z}{c}\right)=\left(1+\frac{y}{b}\right)\left(1-\frac{y}{b}\right).$$

Now let

$$(2)\quad \frac{x}{a}+\frac{z}{c}=m\left(1+\frac{y}{b}\right),\ \frac{x}{a}-\frac{z}{c}=\frac{1}{m}\left(1-\frac{y}{b}\right).$$

For any value of m these equations represent a straight line. It lies on (1) because if the members of (2) are multiplied together (1) results.

But we may also write the equations

$$(3)\quad \frac{x}{a}+\frac{z}{c}=n\left(1-\frac{y}{b}\right),\ \frac{x}{a}-\frac{z}{c}=\frac{1}{n}\left(1+\frac{y}{b}\right).$$

These also represent a line on the surface and thus there are two lines through a given point on the surface that lie wholly on the surface. Each line of either system, (2) or (3), is a generator of the surface.

Similarly the hyperbolic paraboloid (1) of § 131 may be written in the form

$$\left(\frac{x}{a}+\frac{y}{b}\right)\left(\frac{x}{a}-\frac{y}{b}\right)=cz.$$

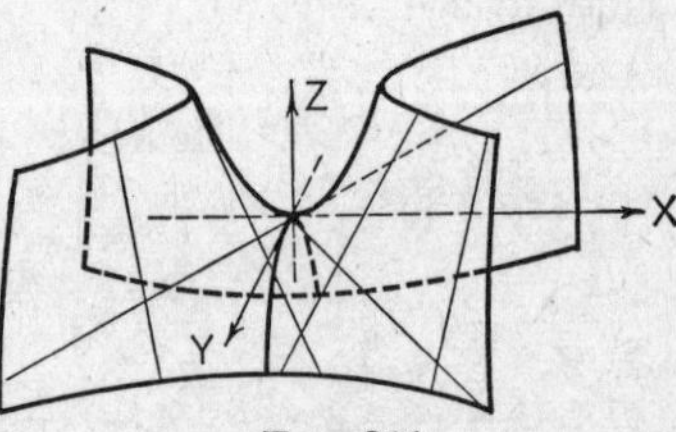

FIG. 211

From this we get the two systems of lines

$$\frac{x}{a}+\frac{y}{b}=mcz,\quad \frac{x}{a}-\frac{y}{b}=\frac{1}{m}z,$$

$$\frac{x}{a}+\frac{y}{b}=\frac{1}{n}z,\quad \frac{x}{a}-\frac{y}{b}=ncz,$$

any member of which lies wholly on the surface and is a generator of the paraboloid.

135. Translations and Rotations. The theory of transformations in three-space is similar to that of transformations in two-space though naturally more complicated in details. (See Chapter X.)

The substitutions

$$x = x' + h,$$
$$y = y' + k,$$
$$z = z' + l$$

will translate the origin from (0, 0, 0) to the point (h, k, l). For an equation with no cross-product terms this translation is

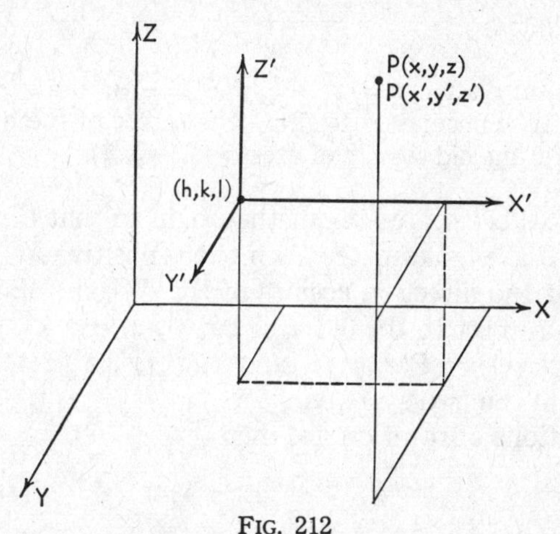

FIG. 212

essentially equivalent to the process of completing the square and will eliminate the non-essential linear terms from the equation

$$(1) \qquad Ax^2 + By^2 + Cz^2 + Gx + Hy + Kz + L = 0.$$

If equation (1) represents a non-central quadric or a parabolic cylinder there is necessarily one linear term which cannot be removed.

Illustration. Simplify the equation $x^2 + 2y^2 - 3z^2 - x + 2y + 9z - 16 = 0$ by removing the linear terms and identify.

Solution. By translating axes to the point (h, k, l) as new origin we get

$$(x'+h)^2+2(y'+k)^2-3(z'+l)^2-(x'+h)+2(y'+k)+9(z'+l)-16=0,$$
$$x'^2+2y'^2-3z'^2+(2h-1)x'+(4k+2)y'+(-6l+9)z'+h^2+$$
$$2k^2-3l^2-h+2k+9l-16=0.$$

To get rid of the linear terms we must take $h = \frac{1}{2}$, $k = -\frac{1}{2}$, $l = \frac{3}{2}$. The reduced equation is

$$x'^2 + 2\,y'^2 - 3\,z'^2 - 10 = 0,$$

$$\frac{x'^2}{10} + \frac{y'^2}{5} - \frac{z'^2}{\frac{10}{3}} = 1. \tag{2}$$

The surface is therefore a hyperboloid of one sheet.

The process of completing the square would yield

$$\frac{(x - \frac{1}{2})^2}{10} + \frac{(y + \frac{1}{2})^2}{5} - \frac{(z - \frac{3}{2})^2}{\frac{10}{3}} = 1$$

and the translation $x = x' + \frac{1}{2}$, $y = y' - \frac{1}{2}$, $z = z' + \frac{3}{2}$ would reduce this immediately to (2). The center of the hyperboloid referred to the old system of axes is $(\frac{1}{2}, -\frac{1}{2}, \frac{3}{2})$.

Let the axes be rotated about the origin so that the positive X-, Y-, and Z-axes go respectively into the positive X'-, Y'-, and Z'-axes. Let the direction cosines of the X'-axis, Y'-axis, and Z'-axis with respect to the old axes be λ_1, μ_1, ν_1; λ_2, μ_2, ν_2; λ_3, μ_3, ν_3 respectively. $P(x, y, z)$ and $P(x', y', z')$ represent the same point in the two systems.

The equations of rotation are then

$$x' = \lambda_1 x + \mu_1 y + \nu_1 z,$$
$$y' = \lambda_2 x + \mu_2 y + \nu_2 z,$$
$$z' = \lambda_3 x + \mu_3 y + \nu_3 z.$$

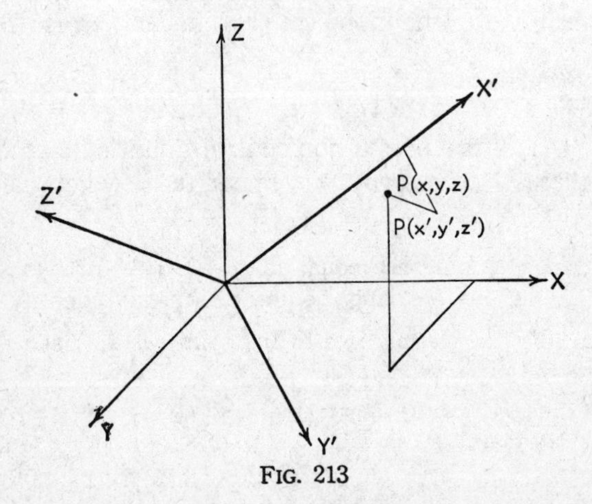

FIG. 213

A proper rotation of axes about the origin will remove the cross-product terms that may be present; a translation will then reduce the equation to one of the standard forms listed in § 133. The complete analysis of the general equation of the second degree is long and involved.

APPENDIX A

Sample Examinations

APPENDIX A

SAMPLE EXAMINATIONS

EXAMINATION I

1. Write the equations of the following lines:

(a) Perpendicular to $x - 2y - 7 = 0$ through $(2, 1)$;

(b) Parallel to the line joining $(1, 3)$, $(0, -2)$ and passing through $(6, -4)$;

(c) Through the intersection of $x - y + 6 = 0$, $3x + 4y - 5 = 0$ and $(-3, 5)$.

2. A triangle has vertices $A(1, 3)$, $B(-2, -1)$, $C(4, -5)$. Find

(a) The area;

(b) The angle CAB.

3. Define a parabola and derive its equation in simplest standard form.

4. Sketch and identify:

(a) $x^2 + 4y^2 - 2x + 24y + 31 = 0$;

(b) $4x^2 - 9y^2 + 18y + 27 = 0$.

5. (a) Find the equation of the hyperbola whose eccentricity is 2 and whose foci are at $(4, 1)$ and $(8, 1)$.

(b) Write the equations of the asymptotes.

6. In polar coordinates sketch $r = 2 - 3 \sin \theta$.

7. By least squares fit a straight line to the following data:

x	0	1	2	3
y	1	2	2	3

8. Write the equations of the following planes:

(a) Perpendicular to the line $\frac{x-1}{2} = \frac{y+3}{1} = \frac{z-5}{-3}$ passing through $(4, 2, -1)$;

(b) Perpendicular to the plane $x - 3y + 3z - 8 = 0$ passing through (1, 1, 0) and (0, 0, 1).

9. Find the direction cosines of the line $x - y + 2z - 1 = 0$, $2x + y + z + 3 = 0$.

10. Identify and sketch the surface $3x^2 - 2y^2 + z^2 - 6x - 8y - 5 = 0$.

EXAMINATION II

1. Given the line $3x + 4y - 20 = 0$. Find the

(a) Intercepts;

(b) Slope;

(c) Distance to (2, −3).

2. Find the locus of a point which moves so that the sum of its distances from (1, 1) and (2, 0) is 5 units.

3. Prove that the line segment joining the midpoints of two sides of a triangle is parallel to the third side and equal to one-half its length.

4. (a) Write the equation of the radical axis of the two circles

$$x^2 + y^2 - 5x + 2y - 8 = 0,$$
$$x^2 + y^2 + 4x - 3y - 1 = 0;$$

(b) Which of the following equations represent parabolas?

$$(1)\ x^2 - y^2 - 2x + y - 1 = 0,$$
$$(2)\ x^2 - y - 2x + y - 1 = 0,$$
$$(3)\ x + y^2 + 2x - y + 1 = 0.$$

5. Write the equations of the following conics:

(a) The hyperbola with vertices at (−2, 0), (4, 0) and conjugate axis 10;

(b) The ellipse whose foci are at (4, 3) and (8, 3) and whose eccentricity is $\frac{1}{2}$.

6. Sketch:

(a) $y = \dfrac{x - 3}{(x - 2)(x + 1)}$;

(b) $r = 2 - \cos\theta$.

7. (a) Show that the curve whose parametric equations are $x = 2t - 5$, $y = 3t + 1$ is a straight line;

(*b*) Transform the polar equation $r^2 = 1 + \sin \theta$ to rectangular coordinates.

8. Sketch:

(a) $y = 2\,e^{-\frac{x}{2}}$;

(b) $y = \sin x + \cos x$.

9. (a) Write the equation of the plane parallel to $3x - y + z - 8 = 0$ and passing through $(1, 2, -2)$;

(b) Find the angle between the planes $x + y - z - 1 = 0$ and $3x + 2y + 3z + 4 = 0$.

10. A point moves so that its distance from the XY-plane is one-half its distance from the origin. Find the equation of the locus and sketch.

EXAMINATION III

1. (a) Derive the equation of the line through the points (x_1, y_1), (x_2, y_2);

(b) Write the equation of the family of lines each member of which is 2 units from the origin.

2. Prove that every angle inscribed in a semicircle is a right angle.

3. (a) Find the equation of the parabola with vertex at $(2, -3)$ and directrix $y = 4$;

(b) What are the coordinates of the focus?

4. (a) Find the equation of the tangent to $\frac{x^2}{9} + \frac{y^2}{4} = 1$ at the point $(2, -\frac{2}{3}\sqrt{5})$;

(b) Sketch $x^2 - \frac{y^2}{4} = 1$ and write down the equations of the asymptotes.

5. Sketch:

(a) $y = \dfrac{x}{(x-1)(x+3)}$;

(b) $y^2 = \dfrac{x}{(x-1)(x+3)}$.

6. (a) Sketch $r = \cos 2\theta$;

(b) Transform $(x - 1)^2 + y^2 = 1$ to polar coordinates.

7. (a) Find the points of intersection of the circle with center at $(3, 1)$ and radius 2 and the X-axis;

(b) Through what angle should the axes be rotated to remove the xy-term from $x^2 - 2\,xy + 3\,y^2 - x + 2 = 0$?

8. (a) Write the equations of the two planes parallel to $4\,x - 2\,y + \sqrt{5}\,z + 1 = 0$ and at a numerical distance of 3 units from it;

(b) Write the equations of the line $x + 2\,y + z - 1 = 0$, $x - y - z + 2 = 0$ in symmetric form.

9. The curve $x^2 + (y - 1)^2 = 1$, $z = 0$ is revolved about the X-axis. Find the equation of the surface generated.

10. A point moves so that the difference between its distances from $(-2, 0, 0)$ and $(2, 0, 0)$ is unity. Find the equation of the locus and identify.

ANSWERS TO EXAMINATIONS

Examination I

1. (a) $2\,x + y - 5 = 0$; (b) $5\,x - y - 34 = 0$; (c) $6\,x + y + 13 = 0$.
2. (a) 18 sq. units; (b) $\tan CAB = \frac{36}{23}$.
3. $y^2 = 4\,px$.
4. (a) Ellipse with center at $(1, -3)$; (b) Hyperbola with center at $(0, 1)$.
5. (a) $\dfrac{(x - 6)^2}{1} - \dfrac{(y - 1)^2}{3} = 1$;
(b) $y = \sqrt{3}\,x + 1 - 6\sqrt{3}$, $y = -\sqrt{3}\,x + 1 + 6\sqrt{3}$.
7. $y = \frac{11}{10} + 3\,x/5$.
8. (a) $2\,x + y - 3\,z - 13 = 0$; (b) $y + z - 1 = 0$.
9. $-\dfrac{1}{\sqrt{3}}, \dfrac{1}{\sqrt{3}}, \dfrac{1}{\sqrt{3}}$.
10. Cone with vertex at $(1, -2, 0)$ and axis perpendicular to the XZ-plane

Examination II

1. (a) $(\frac{20}{3}, 0)$, $(0, 5)$; (b) $m = -\frac{3}{4}$; (c) $-\frac{26}{5}$.
2. $96\,x^2 + 8\,xy + 96\,y^2 - 292\,x - 108\,y - 329 = 0$.
4. (a) $9\,x - 5\,y + 7 = 0$; (b) (2) and (3).
5. (a) $\dfrac{(x - 1)^2}{9} - \dfrac{y^2}{25} = 1$; (b) $\dfrac{(x - 6)^2}{16} + \dfrac{(y - 3)^2}{12} = 1$.
7. (a) Its Cartesian equation is $3\,x - 2\,y + 17 = 0$;
(b) $(x^2 + y^2)(x^2 + y^2 - 1)^2 = y^2$.
9. (a) $3\,x - y + z + 1 = 0$; (b) $\cos\theta = \frac{1}{33}\sqrt{66}$.
10. The cone $x^2 + y^2 - 3\,z^2 = 0$.

Examination III

1. (b) $\lambda x + \mu y - 2 = 0$, $\lambda^2 + \mu^2 = 1$.
3. (a) $(x - 2)^2 = -28(y + 3)$; (b) $(2, -10)$.
4. (a) $4x - 3\sqrt{5}\,y - 18 = 0$; (b) $y = \pm 2x$.
6. (b) $r = 2 \cos \theta$.
7. (a) $(3 \pm \sqrt{3}, 0)$; (b) $22\frac{1}{2}$ degrees.
8. (a) $4x - 2y + \sqrt{5}\,z + 1 \pm 15 = 0$; (b) $\dfrac{x + \frac{1}{2}}{1} = \dfrac{y}{-2} = \dfrac{z - \frac{3}{2}}{3}$.
9. $(x^2 + y^2 + z^2)^2 = 4(y^2 + z^2)$.
10. $60x^2 - 4y^2 - 4z^2 = 15$, hyperboloid of two sheets.

APPENDIX B

Tables

n	n^2	$\sqrt{n}$	n^3	$\sqrt[3]{n}$
1	1	1.000	1	1.000
2	4	1.414	8	1.260
3	9	1.732	27	1.442
4	16	2.000	64	1.587
5	25	2.236	125	1.710
6	36	2.449	216	1.817
7	49	2.646	343	1.913
8	64	2.828	512	2.000
9	81	3.000	729	2.080
10	100	3.162	1,000	2.154
11	121	3.317	1,331	2.224
12	144	3.464	1,728	2.289
13	169	3.606	2,197	2.351
14	196	3.742	2,744	2.410
15	225	3.873	3,375	2.466
16	256	4.000	4,096	2.520
17	289	4.123	4,913	2.571
18	324	4.243	5,832	2.621
19	361	4.359	6,859	2.668
20	400	4.472	8,000	2.714
21	441	4.583	9,261	2.759
22	484	4.690	10,648	2.802
23	529	4.796	12,167	2.844
24	576	4.899	13,824	2.884
25	625	5.000	15,625	2.924
26	676	5.099	17,576	2.962
27	729	5.196	19,683	3.000
28	784	5.291	21,952	3.037
29	841	5.385	24,389	3.072
30	900	5.477	27,000	3.107
31	961	5.568	29,791	3.141
32	1,024	5.657	32,768	3.175
33	1,089	5.745	35,937	3.208
34	1,156	5.831	39,304	3.240
35	1,225	5.916	42,875	3.271
36	1,296	6.000	46,656	3.302
37	1,369	6.083	50,653	3.332
38	1,444	6.164	54,872	3.362
39	1,521	6.245	59,319	3.391
40	1,600	6.325	64,000	3.420
41	1,681	6.403	68,921	3.448
42	1,764	6.481	74,088	3.476
43	1,849	6.557	79,507	3.503
44	1,936	6.633	85,184	3.530
45	2,025	6.708	91,125	3.557
46	2,116	6.782	97,336	3.583
47	2,209	6.856	103,823	3.609
48	2,304	6.928	110,592	3.634
49	2,401	7.000	117,649	3.659
50	2,500	7.071	125,000	3.684
n	n^2	$\sqrt{n}$	n^3	$\sqrt[3]{n}$

n	n^2	$\sqrt{n}$	n^3	$\sqrt[3]{n}$
51	2,601	7.141	132,651	3.708
52	2,704	7.211	140,608	3.732
53	2,809	7.280	148,877	3.756
54	2,916	7.348	157,464	3.780
55	3,025	7.416	166,375	3.803
56	3,136	7.483	175,616	3.826
57	3,249	7.550	185,193	3.848
58	3,364	7.616	195,112	3.871
59	3,481	7.681	205,379	3.893
60	3,600	7.746	216,000	3.915
61	3,721	7.810	226,981	3.936
62	3,844	7.874	238,328	3.958
63	3,969	7.937	250,047	3.979
64	4,096	8.000	262,144	4.000
65	4,225	8.062	274,625	4.021
66	4,356	8.124	287,496	4.041
67	4,489	8.185	300,763	4.062
68	4,624	8.246	314,432	4.082
69	4,761	8.307	328,509	4.102
70	4,900	8.367	343,000	4.121
71	5,041	8.426	357,911	4.141
72	5,184	8.485	373,248	4.160
73	5,329	8.544	389,017	4.179
74	5,476	8.602	405,224	4.198
75	5,625	8.660	421,875	4.217
76	5,776	8.718	438,976	4.236
77	5,929	8.775	456,533	4.254
78	6,084	8.832	474,552	4.273
79	6,241	8.888	493,039	4.291
80	6,400	8.944	512,000	4.309
81	6,561	9.000	531,441	4.327
82	6,724	9.055	551,368	4.344
83	6,889	9.110	571,787	4.362
84	7,056	9.165	592,704	4.380
85	7,225	9.220	614,125	4.397
86	7,396	9.274	636,056	4.414
87	7,569	9.327	658,503	4.431
88	7,744	9.381	681,472	4.448
89	7,921	9.434	704,969	4.465
90	8,100	9.487	729,000	4.481
91	8,281	9.539	753,571	4.498
92	8,464	9.592	778,688	4.514
93	8,649	9.643	804,357	4.531
94	8,836	9.695	830,584	4.547
95	9,025	9.747	857,375	4.563
96	9,216	9.798	884,736	4.579
97	9,409	9.849	912,673	4.595
98	9,604	9.899	941,192	4.610
99	9,801	9.950	970,299	4.626
100	10,000	10.000	1,000,000	4.642
n	n^2	$\sqrt{n}$	n^3	$\sqrt[3]{n}$

N.	0	1	2	3	4	5	6	7	8	9
10	0000	0043	0086	0128	0170	0212	0253	0294	0334	0374
11	0414	0453	0492	0531	0569	0607	0645	0682	0719	0755
12	0792	0828	0864	0899	0934	0969	1004	1038	1072	1106
13	1139	1173	1206	1239	1271	1303	1335	1367	1399	1430
14	1461	1492	1523	1553	1584	1614	1644	1673	1703	1732
15	1761	1790	1818	1847	1875	1903	1931	1959	1987	2014
16	2041	2068	2095	2122	2148	2175	2201	2227	2253	2279
17	2304	2330	2355	2380	2405	2430	2455	2480	2504	2529
18	2553	2577	2601	2625	2648	2672	2695	2718	2742	2765
19	2788	2810	2833	2856	2878	2900	2923	2945	2967	2989
20	3010	3032	3054	3075	3096	3118	3139	3160	3181	3201
21	3222	3243	3263	3284	3304	3324	3345	3365	3385	3404
22	3424	3444	3464	3483	3502	3522	3541	3560	3579	3598
23	3617	3636	3655	3674	3692	3711	3729	3747	3766	3784
24	3802	3820	3838	3856	3874	3892	3909	3927	3945	3962
25	3979	3997	4014	4031	4048	4065	4082	4099	4116	4133
26	4150	4166	4183	4200	4216	4232	4249	4265	4281	4298
27	4314	4330	4346	4362	4378	4393	4409	4425	4440	4456
28	4472	4487	4502	4518	4533	4548	4564	4579	4594	4609
29	4624	4639	4654	4669	4683	4698	4713	4728	4742	4757
30	4771	4786	4800	4814	4829	4843	4857	4871	4886	4900
31	4914	4928	4942	4955	4969	4983	4997	5011	5024	5038
32	5051	5065	5079	5092	5105	5119	5132	5145	5159	5172
33	5185	5198	5211	5224	5237	5250	5263	5276	5289	5302
34	5315	5328	5340	5353	5366	5378	5391	5403	5416	5428
35	5441	5453	5465	5478	5490	5502	5514	5527	5539	5551
36	5563	5575	5587	5599	5611	5623	5635	5647	5658	5670
37	5682	5694	5705	5717	5729	5740	5752	5763	5775	5786
38	5798	5809	5821	5832	5843	5855	5866	5877	5888	5899
39	5911	5922	5933	5944	5955	5966	5977	5988	5999	6010
40	6021	6031	6042	6053	6064	6075	6085	6096	6107	6117
41	6128	6138	6149	6160	6170	6180	6191	6201	6212	6222
42	6232	6243	6253	6263	6274	6284	6294	6304	6314	6325
43	6335	6345	6355	6365	6375	6385	6395	6405	6415	6425
44	6435	6444	6454	6464	6474	6484	6493	6503	6513	6522
45	6532	6542	6551	6561	6571	6580	6590	6599	6609	6618
46	6628	6637	6646	6656	6665	6675	6684	6693	6702	6712
47	6721	6730	6739	6749	6758	6767	6776	6785	6794	6803
48	6812	6821	6830	6839	6848	6857	6866	6875	6884	6893
49	6902	6911	6920	6928	6937	6946	6955	6964	6972	6981
50	6990	6998	7007	7016	7024	7033	7042	7050	7059	7067
51	7076	7084	7093	7101	7110	7118	7126	7135	7143	7152
52	7160	7168	7177	7185	7193	7202	7210	7218	7226	7235
53	7243	7251	7259	7267	7275	7284	7292	7300	7308	7316
54	7324	7332	7340	7348	7356	7364	7372	7380	7388	7396
N.	0	1	2	3	4	5	6	7	8	9

N.	0	1	2	3	4	5	6	7	8	9
55	7404	7412	7419	7427	7435	7443	7451	7459	7466	7474
56	7482	7490	7497	7505	7513	7520	7528	7536	7543	7551
57	7559	7566	7574	7582	7589	7597	7604	7612	7619	7627
58	7634	7642	7649	7657	7664	7672	7679	7686	7694	7701
59	7709	7716	7723	7731	7738	7745	7752	7760	7767	7774
60	7782	7789	7796	7803	7810	7818	7825	7832	7839	7846
61	7853	7860	7868	7875	7882	7889	7896	7903	7910	7917
62	7924	7931	7938	7945	7952	7959	7966	7973	7980	7987
63	7993	8000	8007	8014	8021	8028	8035	8041	8048	8055
64	8062	8069	8075	8082	8089	8096	8102	8109	8116	8122
65	8129	8136	8142	8149	8156	8162	8169	8176	8182	8189
66	8195	8202	8209	8215	8222	8228	8235	8241	8248	8254
67	8261	8267	8274	8280	8287	8293	8299	8306	8312	8319
68	8325	8331	8338	8344	8351	8357	8363	8370	8376	8382
69	8388	8395	8401	8407	8414	8420	8426	8432	8439	8445
70	8451	8457	8463	8470	8476	8482	8488	8494	8500	8506
71	8513	8519	8525	8531	8537	8543	8549	8555	8561	8567
72	8573	8579	8585	8591	8597	8603	8609	8615	8621	8627
73	8633	8639	8645	8651	8657	8663	8669	8675	8681	8686
74	8692	8698	8704	8710	8716	8722	8727	8733	8739	8745
75	8751	8756	8762	8768	8774	8779	8785	8791	8797	8802
76	8808	8814	8820	8825	8831	8837	8842	8848	8854	8859
77	8865	8871	8876	8882	8887	8893	8899	8904	8910	8915
78	8921	8927	8932	8938	8943	8949	8954	8960	8965	8971
79	8976	8982	8987	8993	8998	9004	9009	9015	9020	9025
80	9031	9036	9042	9047	9053	9058	9063	9069	9074	9079
81	9085	9090	9096	9101	9106	9112	9117	9122	9128	9133
82	9138	9143	9149	9154	9159	9165	9170	9175	9180	9186
83	9191	9196	9201	9206	9212	9217	9222	9227	9232	9238
84	9243	9248	9253	9258	9263	9269	9274	9279	9284	9289
85	9294	9299	9304	9309	9315	9320	9325	9330	9335	9340
86	9345	9350	9355	9360	9365	9370	9375	9380	9385	9390
87	9395	9400	9405	9410	9415	9420	9425	9430	9435	9440
88	9445	9450	9455	9460	9465	9469	9474	9479	9484	9489
89	9494	9499	9504	9509	9513	9518	9523	9528	9533	9538
90	9542	9547	9552	9557	9562	9566	9571	9576	9581	9586
91	9590	9595	9600	9605	9609	9614	9619	9624	9628	9633
92	9638	9643	9647	9652	9657	9661	9666	9671	9675	9680
93	9685	9689	9694	9699	9703	9708	9713	9717	9722	9727
94	9731	9736	9741	9745	9750	9754	9759	9763	9768	9773
95	9777	9782	9786	9791	9795	9800	9805	9809	9814	9818
96	9823	9827	9832	9836	9841	9845	9850	9854	9859	9863
97	9868	9872	9877	9881	9886	9890	9894	9899	9903	9908
98	9912	9917	9921	9926	9930	9934	9939	9943	9948	9952
99	9956	9961	9965	9969	9974	9978	9983	9987	9991	9996
N.	0	1	2	3	4	5	6	7	8	9

TABLE III. NATURAL LOGARITHMS (BASE e)

	.00	.01	.02	.03	.04	.05	.06	.07	.08	.09
1.0	0.0000	0.0100	0.0198	0.0296	0.0392	0.0488	0.0583	0.0677	0.0770	0.0862
1.1	0.0953	0.1044	0.1133	0.1222	0.1310	0.1398	0.1484	0.1570	0.1655	0.1740
1.2	0.1823	0.1906	0.1989	0.2070	0.2151	0.2231	0.2311	0.2390	0.2469	0.2546
1.3	0.2624	0.2700	0.2776	0.2852	0.2927	0.3001	0.3075	0.3148	0.3221	0.3293
1.4	0.3365	0.3436	0.3507	0.3577	0.3646	0.3716	0.3784	0.3853	0.3920	0.3988
1.5	0.4055	0.4121	0.4187	0.4253	0.4318	0.4383	0.4447	0.4511	0.4574	0.4637
1.6	0.4700	0.4762	0.4824	0.4886	0.4947	0.5008	0.5068	0.5128	0.5188	0.5247
1.7	0.5306	0.5365	0.5423	0.5481	0.5539	0.5596	0.5653	0.5710	0.5766	0.5822
1.8	0.5878	0.5933	0.5988	0.6043	0.6098	0.6152	0.6206	0.6259	0.6313	0.6366
1.9	0.6419	0.6471	0.6523	0.6575	0.6627	0.6678	0.6729	0.6780	0.6831	0.6881
2.0	0.6932	0.6981	0.7031	0.7080	0.7129	0.7178	0.7227	0.7275	0.7324	0.7372
2.1	0.7419	0.7467	0.7514	0.7561	0.7608	0.7655	0.7701	0.7747	0.7793	0.7839
2.2	0.7885	0.7930	0.7975	0.8020	0.8065	0.8109	0.8154	0.8198	0.8242	0.8286
2.3	0.8329	0.8373	0.8416	0.8459	0.8502	0.8544	0.8587	0.8629	0.8671	0.8713
2.4	0.8755	0.8796	0.8838	0.8879	0.8920	0.8961	0.9002	0.9042	0.9083	0.9123
2.5	0.9163	0.9203	0.9243	0.9282	0.9322	0.9361	0.9400	0.9439	0.9478	0.9517
2.6	0.9555	0.9594	0.9632	0.9670	0.9708	0.9746	0.9783	0.9821	0.9858	0.9895
2.7	0.9933	0.9969	1.0006	1.0043	1.0080	1.0116	1.0152	1.0188	1.0225	1.0260
2.8	1.0296	1.0332	1.0367	1.0403	1.0438	1.0473	1.0508	1.0543	1.0578	1.0613
2.9	1.0647	1.0682	1.0716	1.0750	1.0784	1.0818	1.0852	1.0886	1.0919	1.0953
3.0	1.0986	1.1019	1.1053	1.1086	1.1119	1.1151	1.1184	1.1217	1.1249	1.1282
3.1	1.1314	1.1346	1.1378	1.1410	1.1442	1.1474	1.1506	1.1537	1.1569	1.1600
3.2	1.1632	1.1663	1.1694	1.1725	1.1756	1.1787	1.1817	1.1848	1.1878	1.1909
3.3	1.1939	1.1969	1.2000	1.2030	1.2060	1.2090	1.2119	1.2149	1.2179	1.2208
3.4	1.2238	1.2267	1.2296	1.2326	1.2355	1.2384	1.2413	1.2442	1.2470	1.2499
3.5	1.2528	1.2556	1.2585	1.2613	1.2641	1.2669	1.2698	1.2726	1.2754	1.2782
3.6	1.2809	1.2837	1.2865	1.2892	1.2920	1.2947	1.2975	1.3002	1.3029	1.3056
3.7	1.3083	1.3110	1.3137	1.3164	1.3191	1.3218	1.3244	1.3271	1.3297	1.3324
3.8	1.3350	1.3376	1.3403	1.3429	1.3455	1.3481	1.3507	1.3533	1.3558	1.3584
3.9	1.3610	1.3635	1.3661	1.3686	1.3712	1.3737	1.3762	1.3788	1.3813	1.3838
4.0	1.3863	1.3888	1.3913	1.3938	1.3962	1.3987	1.4012	1.4036	1.4061	1.4085
4.1	1.4110	1.4134	1.4159	1.4183	1.4207	1.4231	1.4255	1.4279	1.4303	1.4327
4.2	1.4351	1.4375	1.4398	1.4422	1.4446	1.4469	1.4493	1.4516	1.4540	1.4563
4.3	1.4586	1.4609	1.4633	1.4656	1.4679	1.4702	1.4725	1.4748	1.4771	1.4793
4.4	1.4816	1.4839	1.4861	1.4884	1.4907	1.4929	1.4951	1.4974	1.4996	1.5019
4.5	1.5041	1.5063	1.5085	1.5107	1.5129	1.5151	1.5173	1.5195	1.5217	1.5239
4.6	1.5261	1.5282	1.5304	1.5326	1.5347	1.5369	1.5390	1.5412	1.5433	1.5454
4.7	1.5476	1.5497	1.5518	1.5539	1.5560	1.5581	1.5602	1.5623	1.5644	1.5665
4.8	1.5686	1.5707	1.5728	1.5748	1.5769	1.5790	1.5810	1.5831	1.5851	1.5872
4.9	1.5892	1.5913	1.5933	1.5953	1.5974	1.5994	1.6014	1.6034	1.6054	1.6074
5.0	1.6094	1.6114	1.6134	1.6154	1.6174	1.6194	1.6214	1.6233	1.6253	1.6273
5.1	1.6292	1.6312	1.6332	1.6351	1.6371	1.6390	1.6409	1.6429	1.6448	1.6467
5.2	1.6487	1.6506	1.6525	1.6544	1.6563	1.6582	1.6601	1.6620	1.6639	1.6658
5.3	1.6677	1.6696	1.6715	1.6734	1.6752	1.6771	1.6790	1.6808	1.6827	1.6845
5.4	1.6864	1.6882	1.6901	1.6919	1.6938	1.6956	1.6974	1.6993	1.7011	1.7029

	.00	.01	.02	.03	.04	.05	.06	.07	.08	.09
5.5	1.7047	1.7066	1.7084	1.7102	1.7120	1.7138	1.7156	1.7174	1.7192	1.7210
5.6	1.7228	1.7246	1.7263	1.7281	1.7299	1.7317	1.7334	1.7352	1.7370	1.7387
5.7	1.7405	1.7422	1.7440	1.7457	1.7475	1.7492	1.7509	1.7527	1.7544	1.7561
5.8	1.7579	1.7596	1.7613	1.7630	1.7647	1.7664	1.7681	1.7699	1.7716	1.7733
5.9	1.7750	1.7766	1.7783	1.7800	1.7817	1.7834	1.7851	1.7868	1.7884	1.7901
6.0	1.7918	1.7934	1.7951	1.7967	1.7984	1.8001	1.8017	1.8034	1.8050	1.8066
6.1	1.8083	1.8099	1.8116	1.8132	1.8148	1.8165	1.8181	1.8197	1.8213	1.8229
6.2	1.8245	1.8262	1.8278	1.8294	1.8310	1.8326	1.8342	1.8358	1.8374	1.8390
6.3	1.8405	1.8421	1.8437	1.8453	1.8469	1.8485	1.8500	1.8516	1.8532	1.8547
6.4	1.8563	1.8579	1.8594	1.8610	1.8625	1.8641	1.8656	1.8672	1.8687	1.8703
6.5	1.8718	1.8733	1.8749	1.8764	1.8779	1.8795	1.8810	1.8825	1.8840	1.8856
6.6	1.8871	1.8886	1.8901	1.8916	1.8931	1.8946	1.8961	1.8976	1.8991	1.9006
6.7	1.9021	1.9036	1.9051	1.9066	1.9081	1.9095	1.9110	1.9125	1.9140	1.9155
6.8	1.9169	1.9184	1.9199	1.9213	1.9228	1.9242	1.9257	1.9272	1.9286	1.9301
6.9	1.9315	1.9330	1.9344	1.9359	1.9373	1.9387	1.9402	1.9416	1.9430	1.9445
7.0	1.9459	1.9473	1.9488	1.9502	1.9516	1.9530	1.9544	1.9559	1.9573	1.9587
7.1	1.9601	1.9615	1.9629	1.9643	1.9657	1.9671	1.9685	1.9699	1.9713	1.9727
7.2	1.9741	1.9755	1.9769	1.9782	1.9796	1.9810	1.9824	1.9838	1.9851	1.9865
7.3	1.9879	1.9892	1.9906	1.9920	1.9933	1.9947	1.9961	1.9974	1.9988	2.0001
7.4	2.0015	2.0028	2.0042	2.0055	2.0069	2.0082	2.0096	2.0109	2.0122	2.0136
7.5	2.0149	2.0162	2.0176	2.0189	2.0202	2.0215	2.0229	2.0242	2.0255	2.0268
7.6	2.0281	2.0295	2.0308	2.0321	2.0334	2.0347	2.0360	2.0373	2.0386	2.0399
7.7	2.0412	2.0425	2.0438	2.0451	2.0464	2.0477	2.0490	2.0503	2.0516	2.0528
7.8	2.0541	2.0554	2.0567	2.0580	2.0592	2.0605	2.0618	2.0631	2.0643	2.0656
7.9	2.0669	2.0681	2.0694	2.0707	2.0719	2.0732	2.0744	2.0757	2.0769	2.0782
8.0	2.0794	2.0807	2.0819	2.0832	2.0844	2.0857	2.0869	2.0882	2.0894	2.0906
8.1	2.0919	2.0931	2.0943	2.0956	2.0968	2.0980	2.0992	2.1005	2.1017	2.1029
8.2	2.1041	2.1054	2.1066	2.1078	2.1090	2.1102	2.1114	2.1126	2.1138	2.1150
8.3	2.1163	2.1175	2.1187	2.1199	2.1211	2.1223	2.1235	2.1247	2.1259	2.1270
8.4	2.1282	2.1294	2.1306	2.1318	2.1330	2.1342	2.1353	2.1365	2.1377	2.1389
8.5	2.1401	2.1412	2.1424	2.1436	2.1448	2.1459	2.1471	2.1483	2.1494	2.1506
8.6	2.1518	2.1529	2.1541	2.1552	2.1564	2.1576	2.1587	2.1599	2.1610	2.1622
8.7	2.1633	2.1645	2.1656	2.1668	2.1679	2.1691	2.1702	2.1713	2.1725	2.1736
8.8	2.1748	2.1759	2.1770	2.1782	2.1793	2.1804	2.1815	2.1827	2.1838	2.1849
8.9	2.1861	2.1872	2.1883	2.1894	2.1905	2.1917	2.1928	2.1939	2.1950	2.1961
9.0	2.1972	2.1983	2.1994	2.2006	2.2017	2.2028	2.2039	2.2050	2.2061	2.2072
9.1	2.2083	2.2094	2.2105	2.2116	2.2127	2.2138	2.2148	2.2159	2.2170	2.2181
9.2	2.2192	2.2203	2.2214	2.2225	2.2235	2.2246	2.2257	2.2268	2.2279	2.2289
9.3	2.2300	2.2311	2.2322	2.2332	2.2343	2.2354	2.2364	2.2375	2.2386	2.2396
9.4	2.2407	2.2418	2.2428	2.2439	2.2450	2.2460	2.2471	2.2481	2.2492	2.2502
9.5	2.2513	2.2523	2.2534	2.2544	2.2555	2.2565	2.2576	2.2586	2.2597	2.2607
9.6	2.2618	2.2628	2.2638	2.2649	2.2659	2.2670	2.2680	2.2690	2.2701	2.2711
9.7	2.2721	2.2732	2.2742	2.2752	2.2762	2.2773	2.2783	2.2793	2.2803	2.2814
9.8	2.2824	2.2834	2.2844	2.2854	2.2865	2.2875	2.2885	2.2895	2.2905	2.2915
9.9	2.2925	2.2935	2.2946	2.2956	2.2966	2.2976	2.2986	2.2996	2.3006	2.3016

N	Nat Log	N	Nat Log	N	Nat Log	N	Nat Log	N	Nat Log
0	—∞	40	3.68 888	80	4.38 203	120	4.78 749	160	5.07 517
1	0.00 000	41	3.71 357	81	4.39 445	121	4.79 579	161	5.08 140
2	0.69 315	42	3.73 767	82	4.40 672	122	4.80 402	162	5.08 760
3	1.09 861	43	3.76 120	83	4.41 884	123	4.81 218	163	5.09 375
4	1.38 629	44	3.78 419	84	4.43 082	124	4.82 028	164	5.09 987
5	1.60 944	45	3.80 666	85	4.44 265	125	4.82 831	165	5.10 595
6	1.79 176	46	3.82 864	86	4.45 435	126	4.83 628	166	5.11 199
7	1.94 591	47	3.85 015	87	4.46 591	127	4.84 419	167	5.11 799
8	2.07 944	48	3.87 120	88	4.47 734	128	4.85 203	168	5.12 396
9	2.19 722	49	3.89 182	89	4.48 864	129	4.85 981	169	5.12 990
10	2.30 259	50	3.91 202	90	4.49 981	130	4.86 753	170	5.13 580
11	2.39 790	51	3.93 183	91	4.51 086	131	4.87 520	171	5.14 166
12	2.48 491	52	3.95 124	92	4.52 179	132	4.88 280	172	5.14 749
13	2.56 495	53	3.97 029	93	4.53 260	133	4.89 035	173	5.15 329
14	2.63 906	54	3.98 898	94	4.54 329	134	4.89 784	174	5.15 906
15	2.70 805	55	4.00 733	95	4.55 388	135	4.90 527	175	5.16 479
16	2.77 259	56	4.02 535	96	4.56 435	136	4.91 265	176	5.17 048
17	2.83 321	57	4.04 305	97	4.57 471	137	4.91 998	177	5.17 615
18	2.89 037	58	4.06 044	98	4.58 497	138	4.92 725	178	5.18 178
19	2.94 444	59	4.07 754	99	4.59 512	139	4.93 447	179	5.18 739
20	2.99 573	60	4.09 434	100	4.60 517	140	4.94 164	180	5.19 296
21	3.04 452	61	4.11 087	101	4.61 512	141	4.94 876	181	5.19 850
22	3.09 104	62	4.12 713	102	4.62 497	142	4.95 583	182	5.20 401
23	3.13 549	63	4.14 313	103	4.63 473	143	4.96 284	183	5.20 949
24	3.17 805	64	4.15 888	104	4.64 439	144	4.96 981	184	5.21 494
25	3.21 888	65	4.17 439	105	4.65 396	145	4.97 673	185	5.22 036
26	3.25 810	66	4.18 965	106	4.66 344	146	4.98 361	186	5.22 575
27	3.29 584	67	4.20 469	107	4.67 283	147	4.99 043	187	5.23 111
28	3.33 220	68	4.21 951	108	4.68 213	148	4.99 721	188	5.23 644
29	3.36 730	69	4.23 411	109	4.69 135	149	5.00 395	189	5.24 175
30	3.40 120	70	4.24 850	110	4.70 048	150	5.01 064	190	5.24 702
31	3.43 399	71	4.26 268	111	4.70 953	151	5.01 728	191	5.25 227
32	3.46 574	72	4.27 667	112	4.71 850	152	5.02 388	192	5.25 750
33	3.49 651	73	4.29 046	113	4.72 739	153	5.03 044	193	5.26 269
34	2.52 636	74	4.30 407	114	4.73 620	154	5.03 695	194	5.26 786
35	3.55 535	75	4.31 749	115	4.74 493	155	5.04 343	195	5.27 300
36	3.58 352	76	4.33 073	116	4.75 359	156	5.04 986	196	5.27 811
37	3.61 092	77	4.34 381	117	4.76 217	157	5.05 625	197	5.28 320
38	3.63 759	78	4.35 671	118	4.77 068	158	5.06 260	198	5.28 827
39	3.66 356	79	4.36 945	119	4.77 912	159	5.06 890	199	5.29 330
40	3.68 888	80	4.38 203	120	4.78 749	160	5.07 517	200	5.29 832

TABLE IV. NATURAL TRIGONOMETRIC FUNCTIONS

Degrees	Radians	Sin	Cos	Tan	Cot	Sec	Csc		
0	.0000	.0000	1.0000	.0000	-----	1.0000	-----	1.5708	90
1	.0175	.0175	.9998	.0175	57.2900	1.0002	57.299	1.5533	89
2	.0349	.0349	.9994	.0349	28.6363	1.0006	28.654	1.5359	88
3	.0524	.0523	.9986	.0524	19.0811	1.0014	19.107	1.5184	87
4	.0698	.0698	.9976	.0699	14.3007	1.0024	14.336	1.5010	86
5	.0873	.0872	.9962	.0875	11.4301	1.0038	11.474	1.4835	85
6	.1047	.1045	.9945	.1051	9.5144	1.0055	9.5668	1.4661	84
7	.1222	.1219	.9925	.1228	8.1443	1.0075	8.2055	1.4486	83
8	.1396	.1392	.9903	.1405	7.1154	1.0098	7.1853	1.4312	82
9	.1571	.1564	.9877	.1584	6.3138	1.0125	6.3925	1.4137	81
10	.1745	.1736	.9848	.1763	5.6713	1.0154	5.7588	1.3963	80
11	.1920	.1908	.9816	.1944	5.1446	1.0187	5.2408	1.3788	79
12	.2094	.2079	.9781	.2126	4.7046	1.0223	4.8097	1.3614	78
13	.2269	.2250	.9744	.2309	4.3315	1.0263	4.4454	1.3439	77
14	.2443	.2419	.9703	.2493	4.0108	1.0306	4.1336	1.3265	76
15	.2618	.2588	.9659	.2679	3.7321	1.0353	3.8637	1.3090	75
16	.2793	.2756	.9613	.2867	3.4874	1.0403	3.6280	1.2915	74
17	.2967	.2924	.9563	.3057	3.2709	1.0457	3.4203	1.2741	73
18	.3142	.3090	.9511	.3249	3.0777	1.0515	3.2361	1.2566	72
19	.3316	.3256	.9455	.3443	2.9042	1.0576	3.0716	1.2392	71
20	.3491	.3420	.9397	.3640	2.7475	1.0642	2.9238	1.2217	70
21	.3665	.3584	.9336	.3839	2.6051	1.0711	2.7904	1.2043	69
22	.3840	.3746	.9272	.4040	2.4751	1.0785	2.6695	1.1868	68
23	.4014	.3907	.9205	.4245	2.3559	1.0864	2.5593	1.1694	67
24	.4189	.4067	.9135	.4452	2.2460	1.0946	2.4586	1.1519	66
25	.4363	.4226	.9063	.4663	2.1445	1.1034	2.3662	1.1345	65
26	.4538	.4384	.8988	.4877	2.0503	1.1126	2.2812	1.1170	64
27	.4712	.4540	.8910	.5095	1.9626	1.1223	2.2027	1.0996	63
28	.4887	.4695	.8829	.5317	1.8807	1.1326	2.1301	1.0821	62
29	.5061	.4848	.8746	.5543	1.8040	1.1434	2.0627	1.0647	61
30	.5236	.5000	.8660	.5774	1.7321	1.1547	2.0000	1.0472	60
31	.5411	.5150	.8572	.6009	1.6643	1.1666	1.9416	1.0297	59
32	.5585	.5299	.8480	.6249	1.6003	1.1792	1.8871	1.0123	58
33	.5760	.5446	.8387	.6494	1.5399	1.1924	1.8361	.9948	57
34	.5934	.5592	.8290	.6745	1.4826	1.2062	1.7883	.9774	56
35	.6109	.5736	.8192	.7002	1.4281	1.2208	1.7434	.9599	55
36	.6283	.5878	.8090	.7265	1.3764	1.2361	1.7013	.9425	54
37	.6458	.6018	.7986	.7536	1.3270	1.2521	1.6616	.9250	53
38	.6632	.6157	.7880	.7813	1.2799	1.2690	1.6243	.9076	52
39	.6807	.6293	.7771	.8098	1.2349	1.2868	1.5890	.8901	51
40	.6981	.6428	.7660	.8391	1.1918	1.3054	1.5557	.8727	50
41	.7156	.6561	.7547	.8693	1.1504	1.3250	1.5243	.8552	49
42	.7330	.6691	.7431	.9004	1.1106	1.3456	1.4945	.8378	48
43	.7505	.6820	.7314	.9325	1.0724	1.3673	1.4663	.8203	47
44	.7679	.6947	.7193	.9657	1.0355	1.3902	1.4396	.8029	46
45	.7854	.7071	.7071	1.0000	1.0000	1.4142	1.4142	.7854	45
		Cos	Sin	Cot	Tan	Csc	Sec	Radians	Degrees

TABLE V. VALUES OF e^x AND e^{-x}

x	e^x	e^{-x}	x	e^x	e^{-x}
0.00	1.0000	1.00000	2.10	8.1662	0.12246
0.01	1.0101	0.99005	2.20	9.0250	0.11080
0.02	1.0202	0.98020	2.30	9.9742	0.10026
0.03	1.0305	0.97045	2.40	11.023	0.09072
0.04	1.0408	0.96079	2.50	12.182	0.08208
0.05	1.0513	0.95123	2.60	13.464	0.07427
0.06	1.0618	0.94176	2.70	14.880	0.06721
0.07	1.0725	0.93239	2.80	16.445	0.06081
0.08	1.0833	0.92312	2.90	18.174	0.05502
0.09	1.0942	0.91393	3.00	20.086	0.04979
0.10	1.1052	0.90484	3.10	22.198	0.04505
0.20	1.2214	0.81873	3.20	24.533	0.04076
0.30	1.3499	0.74082	3.30	27.113	0.03688
0.40	1.4918	0.67032	3.40	29.964	0.03337
0.50	1.6487	0.60653	3.50	33.115	0.03020
0.60	1.8221	0.54881	3.60	36.598	0.02732
0.70	2.0138	0.49659	3.70	40.447	0.02472
0.80	2.2255	0.44933	3.80	44.701	0.02237
0.90	2.4596	0.40657	3.90	49.402	0.02024
1.00	2.7183	0.36788	4.00	54.598	0.01832
1.10	3.0042	0.33287	4.10	60.340	0.01657
1.20	3.3201	0.30119	4.20	66.686	0.01500
1.30	3.6693	0.27253	4.30	73.700	0.01357
1.40	4.0552	0.24660	4.40	81.451	0.01228
1.50	4.4817	0.22313	4.50	90.017	0.01111
1.60	4.9530	0.20190	4.60	99.484	0.01005
1.70	5.4739	0.18268	4.70	109.95	0.00910
1.80	6.0496	0.16530	4.80	121.51	0.00823
1.90	6.6859	0.14957	4.90	134.29	0.00745
2.00	7.3891	0.13534	5.00	148.41	0.00674

INDEX

(The numbers refer to pages.)